M000168754

McGraw-Hill Opitcal and Electro-Optical Engineering Series
Robert E. Fischer and Warren J. Smith, Series Editors

Published

HECHT • *The Laser Guidebook*

MELZER, MOFFITT • *Head Mounted Displays*

SMITH • *Modern Optical Engineering*

SMITH • *Modern Lens Design*

WAYNANT, EDIGER • *Electro-Optics Handbook*

WYATT • *Electro-Optical System Design*

MILLER, FRIEDMAN • *Photonics Rules of Thumb*

Other Books of Interest

OPTICAL SOCIETY OF AMERICA • *Handbook of Optics, Second Edition, Volumes I, II*

KEISER • *Optical Fiber Communications*

SYMS, COZENS • *Optical Waves and Devices*

CHOMYCZ • *Fiber Optical Installations*

To order or receive additional information on these or any other McGraw-Hill titles, in the United States please call 1-800-722-4726. In other countries, contact your local McGraw-Hill representative.

Practical Optical System Layout

And Use of Stock Lenses

Warren J. Smith

Chief Scientist
Kaiser Electro-Optics, Inc.
Carlsbad, California
and Consultant
in Optics and Lens Design

McGraw-Hill

New York San Francisco Washington, D.C. Auckland Bogotá
Caracas Lisbon London Madrid Mexico City Milan
Montreal New Delhi San Juan Singapore
Sydney Tokyo Toronto

Library of Congress Cataloging-in-Publication Data

Smith, Warren J.
    Practical optical system layout / and use of stock lenses / Warren
J. Smith.
        p.   cm.
    Includes index.
    ISBN 0-07-059254-3
    1. Optical instruments—Design and construction. 2. Lenses.
I. Title.
TS513.S56 1997
681'..4—DC21   95-12535
                                                    CIP

McGraw-Hill

A Division of The **McGraw·Hill** Companies

Copyright © 1997 by The McGraw-Hill Companies, Inc. All rights
reserved. Printed in the United States of America. Except as permitted
under the United States Copyright Act of 1976, no part of this publica-
tion may be reproduced or distributed in any form or by any means, or
stored in a data base or retrieval system, without the prior written per-
mission of the publisher.

1 2 3 4 5 6 7 8 9 0 DOC/DOC 9 0 2 1 0 9 8 7

ISBN 0-07-059254-3

*The sponsoring editor for this book was Stephen S. Chapman, the
editing supervisor was Paul R. Sobel, and the production supervisor
was Pamela A. Pelton. It was set in Century Schoolbook by Estelita F.
Green of McGraw-Hill's Professional Book Group composition unit.*

Printed and bound by R. R. Donnelley & Sons Company.

McGraw-Hill books are available at special quantity discounts to use
as premiums and sales promotions, or for use in corporate training programs. For more information, please write to the Director of Special
Sales, McGraw-Hill, 11 West 19th Street, New York, NY 10011. Or
contact your local bookstore.

Information contained in this work has been obtained by The
McGraw-Hill Companies, Inc. ("McGraw-Hill") from sources
believed to be reliable. However, neither McGraw-Hill nor its
authors guarantee the accuracy or completeness of any information published herein, and neither McGraw-Hill nor its authors
shall be responsible for any errors, omissions, or damages aris-
ing out of use of this information. This work is published with
the understanding that McGraw-Hill and its authors are supply-
ing information but are not attempting to render engineering or
other professional services. If such services are required, the
assistance of an appropriate professional should be sought.

This book is printed on recycled, acid-free paper containing
a minimum of 50% recycled, de-inked fiber.

Contents

Preface   ix
Introduction   xi

Chapter 1 The Tools   1

  1.1 Introduction, Assumptions, and Conventions   1
  1.2 The Cardinal (Gauss) Points and Focal Lengths   2
  1.3 The Image and Magnification Equations   4
  1.4 Simple Ray Sketching   12
  1.5 Paraxial Raytracing (Surface by Surface)   17
  1.6 The Thin Lens Concept   22
  1.7 Thin Lens Raytracing   23
  1.8 The Invariant   26
  1.9 Paraxial Ray Characteristics   26
 1.10 Combinations of Two Components   27
 1.11 The Scheimpflug Condition   35
 1.12 Reflectors, Prisms, Mirrors, etc.   37
 1.13 Collected Equations   40

Chapter 2 The Basic Optical Systems   49

  2.1 Introduciton   49
  2.2 Stops and Pupils   49
  2.3 Afocal Systems: General   57
  2.4 Telescopes and Beam Expanders   59
  2.5 Afocal Attachments: Power and Field Changers   66
  2.6 Bravais System   68
  2.7 Field Lenses; Relay Lenses; Periscopes   70
  2.8 Magnifiers and Microscopes   76
  2.9 Telephoto and Retrofocus Arrangements   80
 2.10 Collimators   82
 2.11 Anamorphic Systems   82
 2.12 Zoom and Varifocal Systems   87

2.13 Mirror Systems   88
2.14 Collected Equations   96

Chapter 3 Condensers, Illuminators, Photometry, Etc.   101

3.1 Interchangeability of Sources and Detectors   101
3.2 Koehler Illumination System   101
3.3 Critical Illumination   104
3.4 Illumination Smoothing Devices   105
3.5 Liquid Crystal Display (LCD) and Digital Micromirror Device (DMD
Projectors   106
3.6 Photometry, Radiometry, Illumination, Etc.   110

Chapter 4 System Limits: Performance and Configuration   117

4.1 Introduction   117
4.2 The Diffraction Limit   117
4.3 Image Sensor Limits   121
4.4 Diffraction Limit vs. Sensor Limit   123
4.5 The Optical Invariant   125
4.6 Source and Detector Size Limits   126
4.7 Depth of Focus   127

Chapter 5 How to Lay Out a System   129

5.1 The Process   129
5.2 The Algebraic Approach   131
5.3 The Numerical Solution Method   135
5.4 First-Order Layout by Computer Code   138
5.5 A Quick Rough Sketch   140
5.6 Chromatic Aberration and Achromatism   143
5.7 Athermalization   144
5.8 Sample System Layout   145

Chapter 6 Getting the Most Out of "Stock" Lenses   153

6.1 Introduction   153
6.2 Stock Lenses   153
6.3 Some Simple Measurements   155
6.4 System Mock-up and Test   159
6.5 Aberrations   163
6.6 Capabilities of Various Lens Types   171
6.7 Unusual Lens Types   174
6.8 How to Use a Singlet (Single Element)   178
6.9 How to Use a Cemented Doublet   179

6.10   Combinations of Stock Lenses   180
6.11   Sources   189

Bibliography   193
Index   195

Preface

This book is written to fill the need for a basic text whose subject is limited to the layout of optical systems. It is directed to those who must set up an optical system but who lack either the time or the inclination to probe deeply into geometrical optics and optical design. The book is intended to be a convenient, "stand alone" handbook to which the user can refer in order to find out how to solve a particular optical layout problem.

The process of optical system layout is simply the determination of the optical components (lenses, mirrors, prisms, etc.) as to type and power, and the arrangement or spacing of these components. This must be done in such a way that the system meets its basic function requirements: the production of an image of the desired size and orientation in the desired location, using a system of optics which will fit into the available space, and which will have the capability of providing the required resolution, spot size, MTF, etc. (provided that the lens design of the final system is properly carried out.)

In addition to system designers, potential users of this book might include someone who wants to mock up an optical system as a proof of concept. Another possible user might want a laboratory setup for an experiment or measurement. Yet another might want to develop a system which utilizes stock, off-the-shelf optics, so that small-scale production can begin quickly and without a large up-front investment.

The preliminary design of a complex opitcal system is often carried out, not by a specialist in lens design but by someone who is a systems engineer or a generalist. The configuration of the preliminary optical layout can have a profound and significant effect on the complexity, performance, cost, and tolerance sensitivity of the final "lens-designed" system. Often there exist alternate solutions to a problem which offer significant advantages, yet which may be overlooked.

The first sections of the book are intended to present this subject in a very straightforward, "how-to" way, with little or no theory. The material is presented by both equations and simple diagrams. The intent is that a reader with little or no optics background should be able to pick up the book and immediately put the material to work. The equations are presented succinctly, as "tools," without derivations but with full explanations. Many sample calculations illustrate the process and serve to guide the reader in the use of the "tools."

The central sections of the book illustrate the application of the "tools" to various optical devices such as telescopes, afocal systems, periscopes, magnifiers, and microscopes, and anamorphic and zoom systems. Illumination systems, which are often misunderstood, are explained, and simple expressions are given to allow basic photometric/radiometric calculations to be carried out. The limits imposed on system performance by diffraction, the eye, the sensor, and the through-put invariant are explained. A generalized approach to the solution of a complex system layout is presented, and the relative merits of several ways of executing the approach are discussed. A typical example is presented as an illustration of the system layout process.

The final section of the book attempts to provide the reader with the benefit of a lens designer's knowledge in the form of advice as to how best to use catalog or "stock" optical components in a system, perhaps to test a concept or to prove out a potential product. There are many situations where an experienced lens designer intuitively knows the best way to utilize existing components; this section is intended to pass this type of knowledge along to readers who want to use stock lenses in their work.

This book contains very little material dealing with anything other than the nuts and bolts of optical system layout. Should the reader want to move from the shallow end of the pool to the deep end, the bibliography at the back of the book can provide plenty of material for further reading.

Warren Smith

Introduction

Practical Optical System Layout is intended to help the reader who is undertaking the design of an optical system answer the question, "What do you do when the paper is blank?" This question is often cited by my friend and colleague Max Riedl, who, being of a very orderly and systematic turn of mind, usually proposes as a partial answer, "First you draw the centerline." (Were he more optically and less mechanically inclined, I suspect he might say "optical axis" rather than "centerline.") But behind these slightly cryptic sentences is the implication that every optical system design must begin on a very basic level. The design proceeds from the centerline in orderly steps, by first determining what components are necessary to meet the system requirements. Then the powers of the components and their locations are determined, allowing for any requirements such as clearances, system folds, or packaging constraints. This is optical system layout, and it is the primary topic of this book.

The optical layout process, that is, the choosing of the powers and spacings of the components which make up the optical system, may seem like a trivial thing, since the layout can be completely described in very simple terms, and its development requires only the use of paraxial optics. It is, nonetheless, an extremely important part of the process of developing an optical system, and the simple paraxial layout may play a leading role in determining the cost, complexity, sensitivity, robustness, and performance of the finished system. This is because the layout will determine the speed and angular coverage at which each component is required to perform, and it determines the power of the components and the "work" which each is required to do. A high-powered component contributes heavily to the image-degrading aberrations and must be comprised of many elements in order that these aberrations be corrected. High-powered components are more sensitive to fabrication tolerances and misalignments.

Obviously, more elements and greater sensitivity will result in a temperamental and high-cost system. Often there are several possible system layouts which will meet the specifications of an application; a wise and informed choice among them will pay dividends when the layout is reduced to "glass and brass" (or the modern equivalent).

The layout may be followed by a complete optical- (or lens-) design, or it may be followed by an attempt to model the system using "stock" or "catalog" lenses. Lens design per se is not addressed in this book. However, since some knowledge of the principles of lens design is a definite plus in system layout, we do discuss some aspects which relate to the performance characteristics of lens elements. The selection and use of stock lenses to make up a working system is almost second nature to an experienced lens designer, who knows such things as what sort of lens is capable of a given angular coverage and numerical aperture, and which way a lens should face to get the best results in a given set of circumstances. But to the less experienced, all this can seem an impenetrable jungle of confusion. Chapter 6 attempts to explain such matters in a simple, concise, and straightforward way.

The first chapter is titled "The Tools." It is a collection of the equations which are used regularly in optical system layout work. The significance of each equation and its use are concisely explained; many numerical examples and diagrams accompany each section. No derivations are given; for the reader interested in more depth, an extensive bibliography follows the text. The treatment is limited to paraxial optics, both for the sake of simplicity and because this is all that is necessary in order to do system layout. The usual image-formation equations are given, along with two simple paraxial raytracing schemes. Many optical systems consist of only two major components, and for that reason equations which allow the solution of any and all two-component systems are included in this chapter. A brief discussion of mirrors and prisms, system folding, and image orientation follows. The chapter concludes with a collection of all the equations in one section for easy reference.

In the second chapter the paraxial optics of most of the "standard" optical systems are examined, and equations are presented which allow the system layout to be determined directly from the function requirements. This is done so that the reader will not find it necessary to derive the relationships—there's no point in reinventing the wheel. As in Chap. 1, numerical examples abound. Topics include telescopes, afocal systems, field lenses, relay lenses, periscopes/endoscopes, Bravais systems, microscopes, magnifiers, telephoto systems, retrofocus lenses, collimators, anamorphic systems, zoom lenses, and mirror systems. As in Chap. 1, the equations are collected in the last section of the chapter for quick reference.

The third chapter covers illumination and illuminating systems. The standard illumination systems (Koehler and "critical") are explained, along with the light pipes and beam homogenizer arrays often used to smooth out illumination. The optics of liquid crystal displays, microlenses, and digital micromirror devices are covered, and the chapter concludes with a brief, simplified explanation of photometric or radiometric calculations.

There are many factors which may limit what an optical system can do, and the fourth chapter teaches that it pays to know these limits before undertaking the layout of a system. The well-known effect of diffraction on image definition is a primary limit, and its relationship to the capability of the system's sensor (be it the eye, a detector, or whatever) may be crucial to the ultimate success of the optical system. MTF and depth of focus, their relationship, and the photographic depth of focus are included in this chapter. If the capabilities of the optics are to be fully exploited, there are limits to the smallness of the light source or the detector which can be used. And there is also the optical version of the water-filled-balloon syndrome, where, when you squeeze the beam diameter, you find the beam spread is inexorably increased.

Chapter 5 tells you how to put it all together. Three approaches to optical system layout are presented: the algebraic, the numerical, and the "let the computer do it" approaches. The value of a rough preliminary sketch in evaluating the practicality and potential of a layout is discussed, as is the importance of minimizing the total power of the components of the system. The concepts of althermalizing and achromatism are presented, and finally, an actual system layout, including athermalization and chromatic correction, is demonstrated using the numerical approach.

The final chapter of the book deals extensively with the use of "stock" or "catalog" lenses, discussing their benefits and drawbacks. On the practical side, a number of measurements that can be made without an optical bench, and some ways to mock up a system without fabricating a set of mechanical mounts are presented. In order to make optimum use of a stock lens, one must have some knowledge of its aberration characteristics; these are succinctly presented here so that the aberrations can be recognized and their effects estimated as a function of field and aperture. The capabilities of standard lens assemblies are indicated and the characteristics of some unusual optical elements (aspherics, fresnels, diffractives, index gradients) are presented. Finally, sections covering "how to use—" a singlet element or a cemented doublet to best advantage lead into a description of how to best combine stock elements so as to get the most satisfactory system out of your catalog optics.

The book is fully indexed—indeed, it may be overindexed. Believing that the value of a book as a reference work is greatly enhanced by a strong index, I have elected to err on the side of too long an index, rather than chance a too short one. I hope that this will allow the book to be used as a text, a self-study manual, or reference work once the user has achieved an understanding of the subject.

Practical Optical
System Layout

1

The Tools

1.1 Introduction, Assumptions, and Conventions

This chapter is intended to provide the reader with the tools necessary to determine the location, size, and orientation of the image
formed by an optical system. These tools are the basic paraxial equations which cover the relationships involved. The word "paraxial" is
more or less synonymous with "first-order" and "gaussian"; for our
purposes it means that the equations describe the image-forming
properties of a perfect optical system. You can depend on well-corrected optical systems to closely follow the paraxial laws.

In this book we make use of certain assumptions and conventions
which will simplify matters considerably. Some assumptions will
eliminate a very small minority* of applications from consideration;
this loss will, for most of us, be more than compensated for by a large
gain in simplicity and feasibility.

Conventions and assumptions

1. All surfaces are figures of rotation having a common axis of symmetry, which is called the *optical axis.*

2. All lens elements, objects, and images are immersed in air with
   an index of refraction n of unity.

---

*Primarily, this refers to applications where object space and image space each has a
different index of refraction. The works cited in the bibliography should be consulted in
the event that this or other exceptions to our assumptions are encountered.

3. In the paraxial region *Snell's law* of refraction ($n \sin I = n' \sin I'$) becomes simply $ni = n'i'$, where i and i' are the angles between the ray and the normal to the surface which separates two media whose indices of refraction are n and n'.

4. Light rays ordinarily will be assumed to travel from left to right in an optical medium of positive index. When light travels from right to left, as, for example, after a single reflection, the medium is considered to have a negative index.

5. A distance is considered positive if it is measured to the right of a reference point; it is negative if it is to the left. In Sec. 1.2 and following, the distance to an object or an image may be measured from (a) a focal point, (b) a principal point, or (c) a lens surface, as the reference point.

6. The radius of curvature r of a surface is positive if its center of curvature lies to the right of the surface, negative if the center is to the left. The *curvature c* is the reciprocal of the radius, so that $c = 1/r$.

7. Spacings between surfaces are positive if the next (following) surface is to the right. If the next surface is to the left (as after a reflection), the distance is negative.

8. Heights, object sizes, and image sizes are measured normal to the optical axis and are positive above the axis, negative below.

9. The term "element" refers to a single lens. A "component" may be one or more elements, but it is treated as a unit.

10. The paraxial ray slope angles are not angles but are differential slopes. In the paraxial region the ray "angle" u equals the distance that the ray rises divided by the distance it travels. (It looks like a tangent, but it isn't.)

1.2 The Cardinal (Gauss) Points and Focal Lengths

When we wish to determine the size and location of an image, a complete optical system can be simply and conveniently represented by four axial points called the cardinal, or Gauss, points. This is true for both simple lenses and complex multielement systems. These are the first and second *focal points* and the first and second *principal points*. The focal point is where the image of an infinitely distant axial object is formed. The (imaginary) surface at which the lens appears to bend the rays is called the principal surface. In paraxial optics this surface is called the *principal plane*. The point where the principal plane crosses the optical axis is called the *principal point*.

Figure 1.1 illustrates the Gauss points for a converging lens system. The light rays coming from a distant object at the left define the second focal point F_2 and the second principal point P_2. Rays from an object point at the right define the "first" points F_1 and P_1. The *focal length f* (or effective focal length *efl*) of the system is the distance from the second principal point P_2 to the second focal point F_2. For a lens immersed in air (per assumption 2 in Sec. 1.1), this is the same as the distance from F_1 to P_1. Note that for a converging lens as

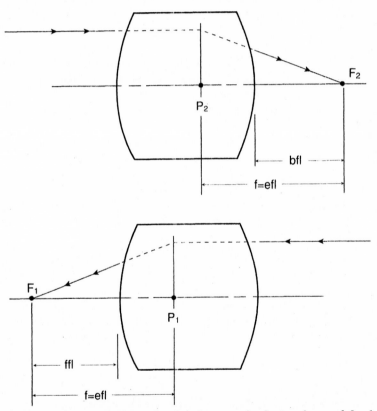

Figure 1.1 The Gauss, or cardinal, points are the first and second focal points F_1 and F_2 and the first and second principal points P_1 and P_2. The focal points are where the images of infinitely distant objects are formed. The distance from the principal point P_2 to the focal point F_2 is the effective focal length *efl* (or simply the focal length f). The distances from the outer surfaces of the lens to the focal points are called the front focal length *ffl* and the back focal length *bfl*.

shown in Fig. 1.1, the focal length has a positive sign according to our sign convention. The *power* ϕ of the system is the reciprocal of the focal length f; $\phi = 1/f$. Power is expressed in units of reciprocal length, e.g., in^{-1} or mm^{-1}; if the unit of length is the meter, then the unit of power is called the *diopter*. For a simple lens which converges (or bends) rays toward the axis, the focal length and power are positive; a diverging lens has a negative focal length and power.

The *back focal length bfl* is the distance from the last (or right-hand) surface of the system to the second focal point F_2. The *front focal length ffl* is the distance from the first (left) surface to the first focal point F_1. In Fig. 1.1, *bfl* is a positive distance and *ffl* is a negative distance. These points and lengths can be calculated by raytracing as described in Sec. 1.5, or, for an existing lens, they can be measured.

The locations of the cardinal points for single-lens elements and mirrors are shown in Fig. 1.2. The left-hand column shows converging, or positive, focal length elements; the right column shows diverging, or negative, elements. Notice that the relative locations of the focal points are different; the second focal point F_2 is to the right for the positive lenses and to the left for the negative. The relative positions of the principal points are the same for both. The surfaces of a positive element tend to be convex and for a negative element concave (exception: a meniscus element, which by definition has one convex and one concave surface, and may have either positive or negative power). Note, however, that a concave mirror acts like a positive, converging element, and a convex mirror like a negative element.

1.3   The Image and Magnification Equations

The use of the Gauss or cardinal points allows the location and size of an image to be determined by very simple equations. There are two commonly used equations for locating an image: (1) Newton's equation, where the object and image locations are specified with reference to the focal points F_1 and F_2, and (2) the Gauss equation, where object and image positions are defined with respect to the principal points P_1 and P_2.

Newton's equation

$$x' = \frac{-f^2}{x} \tag{1.1}$$

where x' gives the image location as the distance from F_2, the second focal point; f is the focal length; and x is the distance from the first

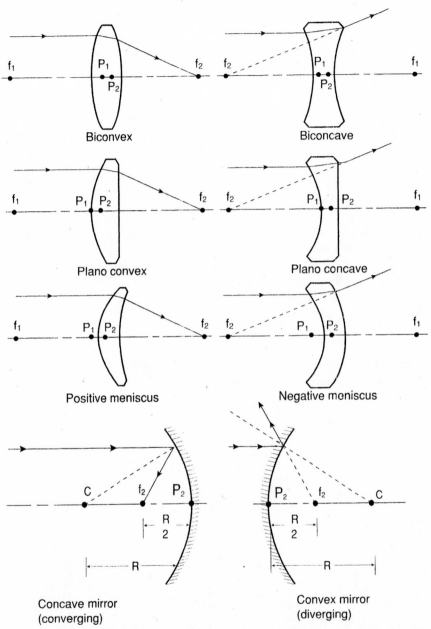

Figure 1.2 Showing the location of the cardinal, or Gauss, points for lens elements. The principal points are separated by approximately $(n-1)/n$ times the axial thickness of the lens. For an equiconvex or equiconcave lens, the principal points are evenly spaced in the lens. For a planoconvex or planoconcave lens, one principal point is always on the curved surface. For a meniscus shape, one principal point is always outside the lens, on the side of the more strongly curved surface. For a mirror, the principal points are on the surface, and the focal length is half of the radius. Note that F_2 is to the right for the positive lens element and to the left for the negative.

focal point F_1 to the object. Given the object size h we can determine the *image size h'* from

$$h' = \frac{hf}{x} = \frac{-hx'}{f} \tag{1.2}$$

The lateral, or transverse, *magnification m* is simply the ratio of image height to object height:

$$m = \frac{h'}{h} = \frac{f}{x} = \frac{-x'}{f} \tag{1.3}$$

Figure 1.3*A* shows a positive focal length system forming a *real image* (i.e., an image which can be formed on a screen, film, CCD, etc.). Note that x in Fig. 1.3*A* is a negative distance and x' is positive; h is positive and h' is negative; the magnification m is thus negative. The image is inverted.

Figure 1.3*B* shows a positive lens forming a *virtual image,* i.e., one which is found inside or "behind" the optics. The virtual image can be seen through the lens but cannot be formed on a screen. Here, x is positive, x' is negative, and since the magnification is positive the image is upright.

Figure 1.3*C* shows a negative focal length system forming a virtual image; x is negative, x' is positive, and the magnification is positive.

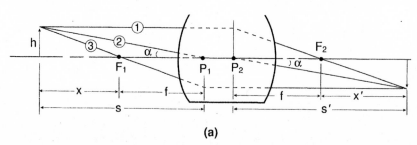

(a)

Figure 1.3 Three examples showing image location by ray sketching and by calculation using the cardinal, or Gauss, points. The three rays which are easily sketched are: (1) a ray from the object point parallel to the axis, which passes through the second focal point F_2 after passing through the lens; (2) a ray aimed at the first principal point P_1, which appears to emerge from the second principal point P_2, making the same angle to the axis α before and after the lens; and (3) a ray through the first focal point F_1, which emerges from the lens parallel to the axis. The distances (s, s', x, and x') used in Eqs. (1.1) through (1.6) are indicated in the figure also. In (*A*) a positive lens forms a real, inverted image.

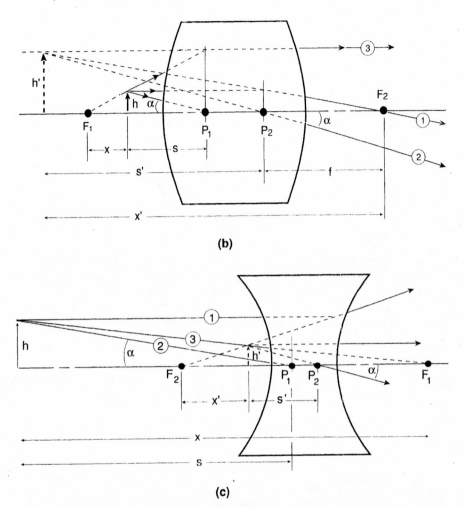

(b)

(c)

Figure 1.3 (*Continued*) Three examples showing image location by ray sketching and by calculation using the cardinal, or Gauss, points. The three rays which are easily sketched are: (1) a ray from the object point parallel to the axis, which passes through the second focal point F_2 after passing through the lens; (2) a ray aimed at the first principal point P_1, which appears to emerge from the second principal point P_2, making the same angle to the axis α before and after the lens; and (3) a ray through the first focal point F_1, which emerges from the lens parallel to the axis. The distances (s, s', x, and x') used in Eqs. (1.1) through (1.6) are indicated in the figure also. In (*B*) a positive lens forms an erect, virtual image to the left of the lens. In (*C*) a negative lens forms an erect, virtual image.

The Gauss equation

$$\frac{1}{s'} = \frac{1}{f} + \frac{1}{s} \tag{1.4}$$

where s' gives the image location as the distance from P_2, the second principal point; f is the focal length; and s is the distance from the first principal point P_1 to the object. The image size is found from

$$h' = \frac{hs'}{s} \tag{1.5}$$

and the transverse magnification is

$$m = \frac{h'}{h} = \frac{s'}{s} \tag{1.6}$$

The sketches in Fig. 1.3 show the Gauss conjugates s and s' as well as the newtonian distances x and x'. In Fig. 1.3A, s is negative and s' is positive. In Fig. 1.3B, s and s' are both negative. In Fig. 1.3C, both s and s' are negative. Note that $s = x-f$ and $s' = x'+f$, and if we neglect the spacing from P_1 to P_2, the object to image distance is equal to $(s-s')$.

Other useful forms of these equations are:

$$s' = \frac{sf}{s + f}$$

$$s' = f(1 - m)$$

$$s = \frac{f(1 - m)}{m}$$

$$f = \frac{ss'}{s - s'}$$

Sample calculations

We will calculate the image location and height for the systems shown in Fig. 1.3, using first the newtonian equations [Eqs. (1.1), (1.2), (1.3)] and then the Gauss equations [Eqs. (1.4), (1.5), (1.6)].
Fig. 1.3A $f = +20, h = +10, x = -25; s = -45$
By Eq. (1.1):

$$x' = \frac{-20^2}{-25} = \frac{-400}{-25} = +16.0$$

Eq. (1.3):

$$m = \frac{20}{-25} = \frac{-16}{20} = -0.8$$

$$h' = -0.8 \cdot 10 = -8.0$$

By Eq. (1.4):

$$\frac{1}{s'} = \frac{1}{20} + \frac{1}{-45} = +0.02777 = \frac{1}{36}$$

$$s' = 36$$

Eq. (1.6):

$$m = \frac{36}{-45} = -0.8$$

$$h' = -0.8 \cdot 10 = -8.0$$

Fig. 1.3*B* $f = +20, h = +10, x = +5; s = -15$

$$x' = \frac{-20^2}{5} = \frac{-400}{5} = -80$$

$$m = \frac{20}{5} = \frac{-(-80)}{20} = +4.0$$

$$h' = 4 \cdot 10 = +40$$

$$\frac{1}{s'} = \frac{1}{20} + \frac{1}{-15} = -0.01666 = \frac{1}{-60}$$

$$s' = -60 \qquad m = \frac{-60}{-15} = +4.0$$

$$h' = 4 \cdot 10 = +40$$

Fig. 1.3*C* $f = -20, h = 10, x = -80; s = -60$

$$x' = \frac{-(-20)^2}{-80} = \frac{-400}{-80} = +5.0$$

$$m = \frac{-20}{-80} = \frac{-5}{-20} = +0.25$$

$$h' = 0.25 \cdot 10 = +2.5$$

$$\frac{1}{s'} = \frac{1}{-20} + \frac{1}{-60} = -0.06666 = \frac{1}{-15}$$

$$s' = -15 \qquad m = \frac{-15}{-60} = +0.25$$

$$h' = 0.25 \cdot 10 = +2.5$$

The image height and magnification equations break down if the object (or image) is at an infinite distance because the magnification becomes either zero or infinite. To handle this situation, we must describe the size of an infinitely distant object (or image) by the angle u_p which it subtends. Note that, for a lens in air, an oblique ray aimed at the first principal point P_1 appears to emerge from the second principal point P_2 with the same slope angle on both sides of the lens. Then, as shown in Fig. 1.4, the image height is given by

$$h' = fu_p \tag{1.7}$$

For trigonometric calculations, we must interpret the paraxial ray slope u as the tangent of the real angle U, and the relationship becomes

$$H' = f \tan U_p \tag{1.8}$$

The *longitudinal magnification M* is the magnification of a dimension *along* the axis. If the corresponding end points of the object and image

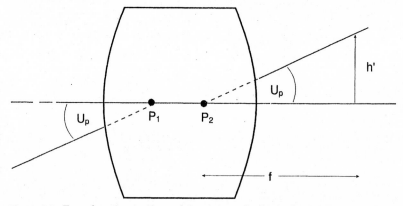

Figure 1.4 For a lens in air, the nodal points and principal points are the same, and an oblique ray aimed at the first nodal/principal point appears to emerge from the second nodal point, making the same angle u_p with the axis as the incident ray. If the object is at infinity, its image is at F_2, and the image height h' is the product of the focal length f and the ray slope (which is u_p for paraxial calculations and $\tan u_p$ for finite angle calculations).

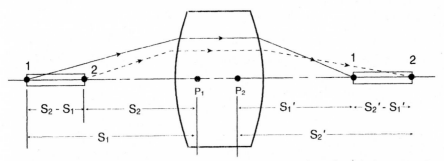

Figure 1.5 Longitudinal magnification is the magnification along the axis (as contrasted to transverse magnification, measured normal to the axis). It can be regarded as the magnification of the thickness or depth of an object or, alternately, as the longitudinal motion of the image relative to that of the object.

are indicated by the subscripts 1 and 2 as shown in Fig. 1.5, then, by definition, the longitudinal magnification is

$$M = \frac{s_2' - s_1'}{s_2 - s_1} \tag{1.9}$$

The value of M can be found from

$$M = \frac{s_1'}{s_1} \cdot \frac{s_2'}{s_2} = m_1 \cdot m_2 \tag{1.10}$$

and in the limit for an infinitesimal length, as m_1 approaches m_2, we get

$$M = m^2 \tag{1.11}$$

Sample calculations

Fig. 1.5 $f = 20, s_1 = -43, s_2 = -39, s_2 - s_1 = 4$
    Exact calculation [Eq. (1.4)]:

$$\frac{1}{s_1'} = \frac{1}{20} + \frac{1}{-43} = 0.02674 = \frac{1}{37.3913}$$

$$\frac{1}{s_2'} = \frac{1}{20} + \frac{1}{-39} = 0.02435 = \frac{1}{41.0526}$$

$$s_2' - s_1' = 41.0526 - 37.3913 = 3.6613$$

Eq. (1.9):

$$M = \frac{3.6613}{4} = +0.91533$$

Note that we can confirm Eq. (1.10):

$$M = m_1 \cdot m_2 = \left(\frac{37.39}{-43.0}\right) \cdot \left(\frac{41.05}{-39.0}\right)$$

$$= (-0.86951)(-1.05263) = +0.91533$$

To use the approximation $M = m^2$, we can take the middle of the object as its location and use -41 for s; then $1/s' = 1/20 + 1/(-41) = 1/39.047619$

$$m = \frac{39.047619}{-41} = -0.952381$$

$$M = m^2 = 0.907029$$

$s_2' - s_1' = 0.907029 \cdot 4 = 3.6281$, approximately the same as the exact value of 3.6613 above.

We can also regard the end points 1 and 2 as different locations for a single object, so that $(s_2 - s_1)$ and $(s_2' - s_1')$ represent the *motion* of object and image. Since Eq. (1.11) shows M equal to a squared quantity, it is apparent that M must be a positive number. The significance of this is that it shows that object and image must both move in the same direction.

Figures 1.6 and 1.7 illustrate this effect for a positive lens (Fig. 1.6) and a negative lens (Fig. 1.7). At the top of each figure the object (solid arrow) is to the left at an infinite distance from the lens and the image is at the second focal point (F_2). As the object moves to the right it can be seen that the image (shown as a dashed arrow) also moves to the right. When the object moves to the first focal point (F_1), the image is then at infinity, which can be considered to be to either the left or the right, and, as the object moves to the right of F_1 the image moves toward the lens, coming from infinity at the left. Note that in the lower sketches the object is projected into the lens and can be considered to be a "virtual object."

1.4 Simple Ray Sketching

When a lens is immersed in air (as per assumption 2 in Sec. 1.1) what are called the *nodal points* coincide with the principal points. The

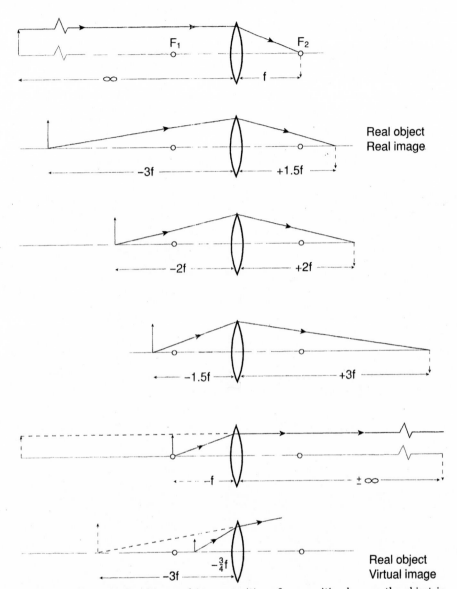

Figure 1.6 Showing the object and image positions for a positive lens as the object is moved from minus infinity to positive infinity.

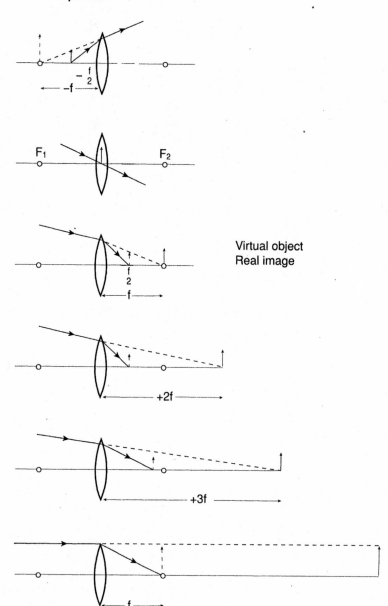

Virtual object
Real image

Figure 1.6 (*Continued*)

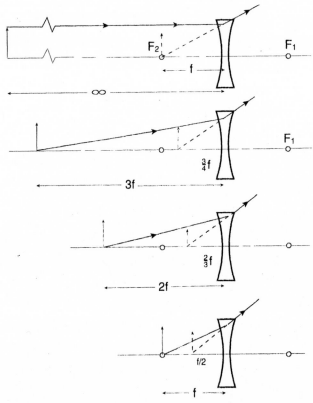

Figure 1.7 Showing the object and image positions for a negative lens as the object is moved from minus infinity to positive infinity.

important characteristic of the nodal points (and thus, for us, the principal points) is that an oblique ray directed toward the first nodal/principal point and making an angle α with the optical axis will emerge (or appear to emerge) from the second nodal/principal point making exactly the same angle α with the axis. [We made use of this in Sec. 1.3, Eqs. (1.7) and (1.8), and in Fig. 1.4.]

There are three rays which can be quickly and easily sketched to determine the image of an off-axis object point. The first ray is drawn from the off-axis point, parallel to the optical axis. This ray will be bent (or *appear* to be bent) at the second principal plane and then must pass through the second focal point. The second ray is the nodal point ray described above. The third is a ray directed toward the first focal point, which is bent at the first principal plane and emerges par-

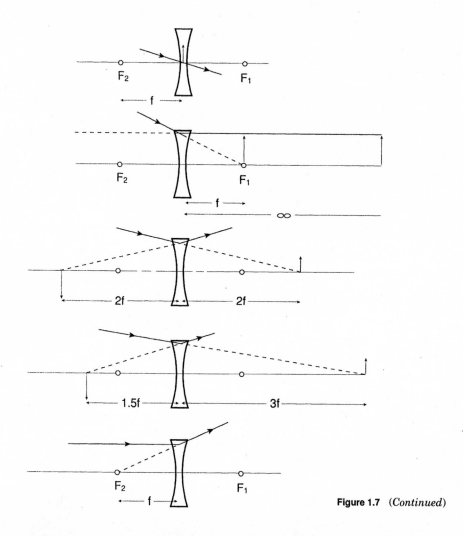

Figure 1.7 (*Continued*)

allel to the axis. This third ray is simply the reverse direction version of the ray which defines the first focal and principal points.

The intersection of any two of these rays serves to locate the image of the object point. If we repeat the exercise for a series of object points along a straight object line perpendicular to the axis, its image will be found to be a straight line also perpendicular to the axis. Thus when the location of one image point has been determined, the image of the entire object line is known. Three examples of this ray sketching technique are illustrated in Fig. 1.3. Note also that this technique can be applied, one lens at a time, to a series of lenses in order to determine the imagery of a complex system, although the process could become quite tedious for a complex system.

If the object plane is tilted, the image tilt can be determined using the Scheimpflug condition, which is described in Sec. 1.11.

1.5 Paraxial Raytracing (Surface by Surface)

This section is presented primarily for reference purposes. This material is not necessary for optical layout work; however, it is included for completeness (and for the benefit of those whose innate curiosity may cause them to wonder just how focal lengths, cardinal points, etc., are determined).

The focal lengths and the cardinal points can be calculated by tracing ray paths, using the equations of this section. The rays which are traced are those which we used in defining the cardinal points in Sec. 1.2. Here, we assume that the construction data of the lens system (radius r, spacing t, and index n) are known and that we desire to determine either the cardinal points and focal lengths or, alternately, to determine the image size and location for a given object.

A ray is defined by its slope u and the height y at which it strikes a surface. Given y and u, the distance l to the point at which the ray crosses, or intersects, the optical axis is given by

$$l = \frac{-y}{u} \tag{1.12}$$

If u is the slope of the ray before it is refracted (or reflected) by a surface and u' is the slope after refraction, then the intersection length after refraction is

$$l' = \frac{-y}{u'} \tag{1.13}$$

The intersection of two (or more) rays which originate at an object point can be used to locate the image of that point. If we realize that the optical axis is in fact a ray, then a single ray starting at the foot (or axial intersection) of an object can be used to locate the image, simply by determining where that ray crosses the axis after passing through the optical system. Thus l and l' above can be considered to be object and image distances, as illustrated in Fig. 1.8.

The raytracing problem is simply this: Given a surface defined by r, n, and n', plus y and u to define a ray, find u' after refraction. The equation for this is

$$n'u' = nu - \frac{y(n' - n)}{r} \tag{1.14a}$$

$$n'u' = nu - \frac{y(n' - n)}{c} \tag{1.14b}$$

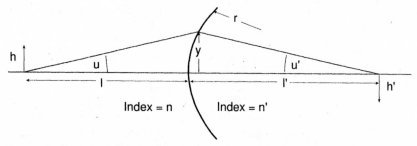

Figure 1.8 A ray traced through a single surface, with the dimensions used in the raytrace calculation labeled for identification.

where n and n' are the indices of refraction for the materials before and after the surface, r is the radius of the surface, c is the curvature of the surface and is the reciprocal of $r(c = 1/r)$, u and u' are the ray slopes before and after the surface, and y is the height at which the ray strikes the surface.

The sign convention here is that a ray sloping upward to the right has a positive slope (as u in Fig. 1.8) and a ray sloping downward (as u' in Fig. 1.8) has a negative slope. A radius has a positive sign if its center of curvature is to the right of the surface (as in Fig. 1.8). A ray height above the axis is positive (as y in Fig. 1.8). The distances l and l' are positive if the ray intersection point is to the right of the surface (as is l' in Fig. 1.5) and negative if to the left (as is l in Fig. 1.5).

In passing we can note that the term

$$\frac{n' - n}{r} = (n' - n)c$$

is called the *surface power*; if r is in meters, the unit of power is the *diopter*.

Sample calculation

Fig. 1.8 $l = -100; y = 10$
Eq. (1.12):

$$u = \frac{-10}{-100} = +0.10$$

$$n = 1.0 \qquad r = +20 \qquad n' = 2.0$$

Eq. (1.14):

$$n'u' = 1.0 \cdot 0.10 - \frac{10(2.0-1.0)}{20}$$

$$= 0.1 - 0.5 = -0.4$$

$$u' = \frac{n'u'}{n'} = \frac{-0.4}{2.0} = -0.2$$

Eq. (1.13):

$$l' = \frac{-10}{-0.2} = +50$$

Looking ahead to Eq. (1.16a), we can also get

$$m = \frac{nu}{n'u'} = \frac{1 \cdot 0.1}{2 \cdot (-0.2)} = -0.25$$

To trace a ray through a series of surfaces, we also need an equation to transfer the ray from one surface to the next. The height of the ray at the next surface (call it y_2) is equal to the height at the current surface (y_1) less the amount the ray drops traveling to the next surface. If the distance measured along the axis to the next surface is t, the ray will drop $(-tu')$ and

$$y_2 = y_1 + tu_1' \tag{1.15}$$

To trace the path of a ray through a system of several surfaces, we simply apply Eqs. (1.14) and (1.15) iteratively throughout the system, beginning with y_1 and u_1, until we have the ray height y_k at the last (or kth) surface and the slope u_k' after the last surface. Then Eq. (1.13) is used to determine the final intersection length l_k'. If the axial intersection points of the ray represent the object and image locations, then the system magnification can be determined from

$$m = \frac{h'}{h} = \frac{n_1 u_1}{n_k' u_k'} \tag{1.16a}$$

or, since we have assumed that for our systems both object and image are in air of index $n = 1.0$,

$$m = \frac{h'}{h} = \frac{u_1}{u_k'} \tag{1.16b}$$

If we want to determine the cardinal points of an air-immersed system, we start a ray parallel to the axis with $u_1 = 0.0$ (as shown in Fig. 1.9)

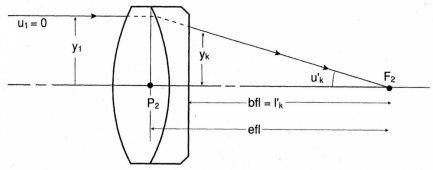

Figure 1.9 The calculation of the cardinal points is done by tracing the path of a ray with an initial slope of zero. The second focal point F_2 is at the final intersection of this ray with the axis; the back focal length is $-y_k/u_k'$ and the effective focal length is equal to $-y_1/u_k'$.

and with y_1 at any convenient value. Then the focal length and back focal length are given by

$$f = efl = \frac{-y_1}{u_k'} \tag{1.17}$$

$$bfl = \frac{-y_k}{u_k'} \tag{1.18}$$

The back focal length bfl obviously locates the second focal point F_2, and $(bfl-efl)$ is the distance from the last surface to the second principal point P_2. The "first" points P_1 and F_1 can be determined by reversing the lens and repeating the process. Note that since the paraxial equations in general, and these equations in particular, are linear in y and u, we get exactly the same results for efl and bfl regardless of the initial value we select for y_1.

Sample calculations

Fig. 1.9   Given the lens data in the following tabulation, find the efl and bfl. Use Eqs. (1.14), (1.15), (1.17), and (1.18).

Surface no.		1		2		3	
Radius		+50		−50		∞	
Curvature		+0.02		−0.02		0.0	
Spacing			6.0		3.0		
Index	1.0		1.5		1.6		1.0
y =		10		9.60		9.4485	
nu =	0.0		−0.10		−0.0808		−0.0808
u =	0.0		−0.0666		−0.0505		−0.0808

$$efl = \frac{-y_1}{u_k{}'} = \frac{-10}{-0.0808} = +123.762376$$

$$bfl = \frac{-y_k{}'}{u_k{}'} = \frac{-9.4485}{-0.0808} = +116.936881$$

$$bfl - efl = -6.825495$$

If the object at infinity subtends a paraxial angle of 0.1 or a real angle whose tangent is 0.1, the image size is $h' = u_p \cdot f = 0.1 \cdot 123.762 = 12.3762$ per Eq. (1.7) or (1.8).

Reversing the lens and repeating the calculation, we get exactly the same focal length (this makes a good check on our calculation) and find that the $ffl = -124.752475$ and that P_1 is $(ffl + efl) = -0.990099$ (to the left) from surface 1.

If we want to determine the imagery of an object 15 mm high which is located 500 mm to the left of the first surface of this lens, we can locate the image by tracing a ray from the foot of the object and determining where the ray crosses the axis after passing through the lens. We use a ray height y of 10 at surface 1; then $u_1 = 10/500 = +0.02$ and the raytrace data is

Surface no.		1		2	3		
$y =$		10		9.68		9.5663	
$nu =$	+0.020		−0.08		−0.06064		−0.06064
$u =$	+0.020		−0.05333		−0.03790		−0.06064

$$l_3{}' = \frac{-y_3}{u_3{}'} = \frac{-9.5663}{-0.06064} = +157.755607$$

$$m = \frac{u_1}{u_3{}'} = \frac{0.02}{-0.06064} = -0.329815$$

$$h' = m \cdot h = -0.3298 \cdot 15 = -4.947230$$

Since we calculated the cardinal points and the focal lengths in the first part of this example, we could also use Eqs. (1.1) and (1.4) to locate the image, as follows:

This object is $500 - 0.990099 = 499.009901 = -s$ to the left of P_1 and $500 - 124.752475 = 375.247525 = -x$ to the left of F_1.

So by Newton [Eq. (1.1)], $x' = -(123.762)^2/(-375.247) = +40.818725$ from F_2, or $40.818 + 116.939 = 157.755607$ from surface 3. The magnification [Eq. (1.3)] is $m = 123.762/(-375.247) = -0.329815$.

And by Gauss [Eq. (1.4)], $1/s' = 1/123.762 + 1/(-499.0099) = 0.006076$ or 164.581102 from P_2 and $164.581 - 6.625 = 157.755643$ from surface 3. The magnification [Eq. (1.6)] is $m = 164.581/(499.0099) = -0.329815$.

This demonstrates that all the equations given above will yield exactly the same answer.

A note about _paraxial._ The reader may be aware that the paraxial equations that we are using are actually differential expressions and thus are rigorously valid only for infinitesimal angles and ray heights. The paraxial region has been well described as a "thin, threadlike region about the optical axis." Yet in our raytracing we are using real, finite ray heights and slopes (angles). The paraxial equations actually represent the relationships or ratios of the various quantities as they approach zero as a limit. These ratios and the (limiting) axial intersection lengths are perfectly exact. However, the ray heights and angles of a paraxial raytrace are not the same values that one would obtain from an exact, trigonometric raytrace of the same starting ray using Snell's law. So in one sense the paraxial raytrace is perfectly exact, justifying a precision to as many decimal places as are useful, while in another sense it is also correct to refer to "the paraxial approximation."

1.6   The Thin Lens Concept

A _thin lens_ is simply one whose axial thickness is zero. Obviously no _real_ lens has a zero thickness. The thin lens is a _concept_ which is an extremely useful tool in optical system layout, and we make extensive use of it in this book. When a lens or optical system has a zero thickness, the object and image calculations can be greatly simplified. In dealing with a thick lens we must be concerned with the location of the principal points. But if the lens has a zero thickness, the two principal points are coincident and are located where the lens is located. Thus by using the thin lens concept in our system layout work, we need only consider the location and power of the component. We can represent a lens of any degree of complexity as a thin lens. What in the final system may be a singlet, doublet, triplet, or a complex lens of 10 or 15 elements may be treated as a thin lens, when our concern is to determine the size, orientation, and location of the image.

The drawbacks to using the thin lens concept are that a thin lens is not a real lens, that its utility is limited to the paraxial region, and that we must ultimately convert the thin lens system into real lenses with radii, thicknesses, and materials. However, for the layout of optical systems, the thin lens concept is of unsurpassed utility. The replacement of thin lenses with thick lenses is actually quite easy. The complete description of a thin lens system consists of just a set of lens powers and the spacings between the lenses. The corresponding thick lens system must have the same component powers (or focal lengths) and the components must be spaced apart by the same spacings, but the thick lens spacings must be measured from the principal points of the lenses. For example, if we have a two-component system with a thin lens spacing of 100 mm, then our thick lens system must have two components whose spacing is also 100 mm, but the 100 mm is measured from the second principal point (P_2) of the first component to the first principal point (P_1) of the second component. The result is a real system which has exactly the same image size, orientation, and location as the thin lens system.

1.7   Thin Lens Raytracing

Because of the words "thin lens" in the title of this section, a rereading of Sec. 1.6 may be needed to remind us that we can also use the raytracing of this section for thick, complex components, provided that we use ray heights at, and spacings from, their principal planes. However, our primary concern here is to provide a very simple raytracing scheme for us to use in optical system layout work, and to this end we deal primarily with the "thin lens" concept.

The raytracing equations that we use are very simple and easy to remember. To trace a ray through a lens we use

$$u' = u - y\phi \tag{1.19}$$

where u' is the ray slope after passing through the lens, u is the ray slope before the lens, y is the height at which the ray strikes the lens (or appears to strike the principal planes), and ϕ is the power of the lens, equal to the reciprocal of the focal length. To transfer the ray to the next lens of a system, we use

$$y_{j+1} = y_j + du_j' \tag{1.20}$$

where y_{j+1} is the ray height at the next [(j+1)th] lens, y_j is the ray height at the current (jth) lens, d is the distance or space between the lenses, and u_j' is the ray slope after passing through the jth element.

Note that when these equations are used for thick components, the ray heights are at the principal planes (that is, they indicate where the ray in air, if extended, would intersect the principal plane), the ray height at the first principal plane is identical to that at the second principal plane, and the spacing d in Eq. (1.20) is measured from the second principal plane of the *j*th component to the first principal plane of the (*j*+1)th component.

Sample calculation

Given three lenses with powers of + 0.01, −0.022, and + 0.022, with spacings of 15.0 between them as shown in Fig. 1.10, we want to find the *efl* and *efl* of the combination. We start by tracing the focal length defining ray with a slope u_1 of zero, with a convenient ray height y_1 of 10, and applying first Eq. (1.19), then Eq. (1.20), then Eq. (1.19), etc.

Lens no.		1		2		3	
Power		+ 0.010		−0.022		+ 0.022	
Spacing			15		15		
y =		10.0		8.5		9.805	
u =	0.00		−0.10		+0.087		−0.12871

Then we can get the focal length from Eq. (1.17):

$$f = efl = \frac{-y_1}{u_k'}$$

$$= \frac{-10.0}{-0.12871} = +77.694$$

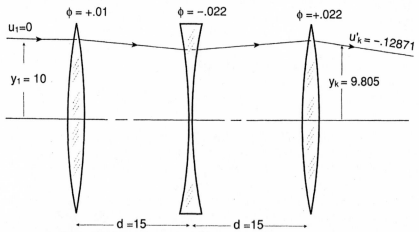

Figure 1.10 The thin lens system used in the sample calculation, with initial and final ray data for the focal length calculation.

and we can get the back focal length from Eq. (1.18) [or from Eq. (1.13)]:

$$bfl = \frac{-y_k}{u_k{}'}$$

$$= \frac{-9.805}{-0.12871} = +76.179$$

Sample calculation

If an infinitely distant object subtends an angle of 0.15 from the system of our previous example, we find the image height from Eq. (1.7):

$$h' = fu_p = 77.694 \cdot 0.15 = 11.654$$

Sample calculation

If our system has a finite object 500 mm away and 15 mm high, one of the methods we can use to determine the image size and location is to trace a ray from the foot of the object. If we use a ray which strikes the first lens at a height y_1 of 10.0, it will have a slope in object space of $u_1 = 10/500 = +0.02$ and our raytrace data is tabulated as follows:

Lens no.		1		2		3	
Power		+ 0.01		−0.022		+ 0.022	
Spacing	500		15		15		
$y =$		10.0		8.80		10.504	
$u =$	+ 0.02		−0.08		+ 0.1136		−0.117488

From this data we determine the image position

$$l_k{}' = \frac{-y_k}{u_k{}'}$$

$$= \frac{-10.504}{-0.117488} = 89.4049$$

and the magnification and image size

$$m = \frac{u_1}{u_k{}'}$$

$$= \frac{0.02}{-0.117488} = -0.170230$$

$$h' = mh = -0.170230 \cdot 15 = -2.55345$$

Note that negative magnification and the negative image height indicate that the image is inverted.

1.8 The Invariant

An algebraic combination of raytracing quantities which is exactly the same anywhere in an optical system is called an invariant. The *Lagrange invariant* has significant utility in optical system layout. It gives the relationship between two paraxial rays, which are usually (but not necessarily) an *axial* ray (from the center of the object through the edge of the lens aperture) and a *chief* or *principal* ray (from the edge of the object through the center of the lens aperture). The invariant can be written

$$\mathrm{INV} = n(y_p u - y u_p) = n'(y_p u' - y u_p') \tag{1.21}$$

where n is the index, u and y are the slope and ray height of the axial ray, and u_p and y_p are the slope and height of the principal ray. This expression can be evaluated for a given pair of rays before or after any surface or lens of a system, and will have exactly the same value at *any other surface or lens* of the system.

If, for example, we evaluate the invariant at the object surface, then y for the axial ray must be zero (by definition) and y_p is the object height h; the invariant becomes $\mathrm{INV} = hnu$. However, at the image surface it must equal $h'n'u'$, and we have

$$\mathrm{INV} = hnu = h'n'u' \tag{1.22}$$

which indicates the immutable relationship between image height h and convergence angle u, and also, confirming Eq. (1.16), our expression for magnification.

The square of the invariant $h^2 n^2 u^2$ is obviously related to the product of the object area and the solid angle of the acceptance cone of rays (or the image area and the cone of illumination). It can also be applied to the area of the entrance or exit pupil (see Sec. 2.2) and the solid angle of the field of view. It is thus a measure of the ability of the optical system to transmit power or information. This illustrates the concept behind the terms "throughput" and "etendue"; obviously the square of the invariant is also invariant.

1.9 Paraxial Ray Characteristics

Because the paraxial raytracing equations are simple linear expressions in y and u, any paraxial raytrace may be scaled. That is, all of the ray data may be multiplied (or divided) by the same constant, and the result will be the raytrace data for a new ray. This new ray will have the same axial intersection locations as the original, but all the ray slopes and heights will be scaled by the same factor. We could, for example, take the sample calculation raytrace of Sec. 1.7 and scale it by a factor of 1.5. The starting ray height would then be $1.5 \cdot 10.0 = 15.0$ and the final ray slope would be $1.5 \cdot (-0.07129) = -0.106935$. Note well, however, that the focal length and the back focal length derived from this new ray data will be exactly the same as those from the original ray data.

Another interesting aspect of paraxial rays occurs if we consider the ray data simply as a set of equalities, as, for example, $u' = u - y\phi$, or $y_2 = y_1 + du_1'$, and realize that when we add equalities, the result is still an equality. Thus if we add (or subtract) the data of two rays, we get the data of yet a third ray. This new data is a perfectly valid description of another ray. Suppose we have traced two rays, 1 and 2. We can scale these rays by multiplying their data by constants A and B to get sets of ray data Ay_1, Au_1 and By_2, Bu_2. Now we can add the scaled ray data to create ray 3 as follows:

$$u_3 = Au_1 + Bu_2 \qquad (1.23)$$

$$y_3 = Ay_1 + By_2 \qquad (1.24)$$

When we want to create a specific ray 3, we need to know the appropriate values of the scaling constants A and B in order to do so. If we know the data of the third ray at some location in the system where we also know the data of rays 1 and 2, then the required scaling factors can be found from

$$A = \frac{y_3 u_1 - u_3 y_1}{u_1 y_2 - y_1 u_2} \qquad (1.25)$$

$$B = \frac{u_3 y_2 - y_3 u_2}{u_1 y_2 - y_1 u_2} \qquad (1.26)$$

In practice ray 1 is often the axial ray and ray 2 is the principal ray, as cited in Sec. 1.8.

1.10   Combinations of Two Components

A great many optical systems consist of just two separated components. The following expressions can be used to handle any and all two-component systems. Note that these equations are valid for thick components as well as "thin lenses." Also, these expressions are valid for components of any degree of complexity. For thick lenses the spacings are measured from their principal points. For thin lenses the spacings are simply the lens-to-lens distances.

Given.   The powers (or focal lengths) of the components and their spacing.

Find.   The ef_1, bf_1, and ff_1 of the combination. See Fig. 1.11.
Power:

$$\phi_{ab} = \phi_a + \phi_b - d\phi_a\phi_b \tag{1.27}$$

efl:

$$f_{ab} = \frac{f_a f_b}{f_a + f_b - d} \tag{1.28}$$

bfl:

$$B = \frac{f_{ab}(f_a - d)}{f_a} \tag{1.29}$$

ffl:

$$FF = \frac{-f_{ab}(f_b - d)}{f_b} \tag{1.30}$$

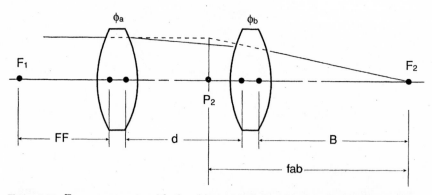

Figure 1.11   Two components with the object at infinity, showing the spacing, the second principal point, the effective focal length, the back focus distance *B*, and the front focus distance *FF.*

Given. The *efl*, *d*, and *B* of the combination.

Find. The focal lengths or powers of the components. See Fig. 1.11.

$$f_a = \frac{df_{ab}}{f_{ab} - B} = \frac{1}{\phi_a} \qquad (1.31)$$

$$f_b = \frac{-dB}{f_{ab} - B - d} = \frac{1}{\phi_b} \qquad (1.32)$$

Finite conjugate systems. See Fig. 1.12.

Given. The component locations (defined by the object distance s, the image distance s', and the spacing d) and the magnification $m = h'/h = u/u'$.

Find. The component powers.

$$\phi_a = \frac{ms - md - s'}{msd} \qquad (1.33)$$

$$\phi_b = \frac{d - ms + s'}{ds'} \qquad (1.34)$$

Given. The component powers, the object-to-image distance, and the magnification $m = h'/h = u/u'$.

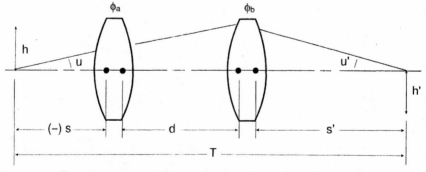

Figure 1.12 Two components working at finite conjugate distances, showing the spacing and object and image distances.

Find. The component locations (i.e., s, s', and d). Solve this quadratic for d:

$$0 = d^2 - dT + T(f_a + f_b) + \frac{(m-1)^2 f_a f_b}{m} \tag{1.35}$$

using
$$x = \frac{-b \pm \sqrt{b^2 - 4ac}}{2a}$$

Then:

$$s = \frac{(m-1)d + T}{(m-1) - md\phi_a} \tag{1.36}$$

$$s' = T + s - d \tag{1.37}$$

Sample calculation

Find the Gauss points of a system whose first lens has a focal length of 200 mm and whose second lens has a focal length of 100 mm, where the separation between lenses is 100 mm. See Fig. 1.13. Power by Eq. (1.27):

$$\phi_{ab} = 0.005 + 0.01 - 100 \cdot 0.005 \cdot 0.01 = 0.01 \quad \left(f = \frac{1}{\phi} = 100.0 \right)$$

Focal length by Eq. (1.28):

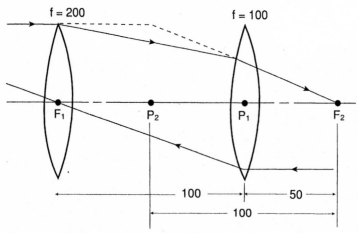

Figure 1.13 Showing the system used for the sample calculation of the cardinal points of a two component system.

$$f_{ab} = \frac{200 \cdot 100}{200 + 100 - 100} = 100.0$$

Back focus by Eq. (1.29):

$$B = \frac{100(200 - 100)}{200} = 50.0$$

Front focus by Eq. (1.30):

$$FF = \frac{-100(100 - 100)}{100} = 0.0$$

F_1 is at the front lens ($FF = 0.0$).

P_1 is at the rear lens, and 100 mm to the right of the front lens.

F_2 is 50 mm to the right of the rear lens ($B = 50.0$).

P_2 is 50 mm to the right of the front lens, and 50 mm to the left of the rear lens.

Sample calculation

Find the component powers necessary to produce an *erect* image 15 mm high from a distant object which subtends an angle of 0.01. The system should be 250 mm long from front lens to image. See Fig. 1.14.

For an erect image the focal length must be negative. [Consider Eqs. (1.7) and (1.8).] Its magnitude [from Eq. (1.7)] is $15.0/0.01 = 150$

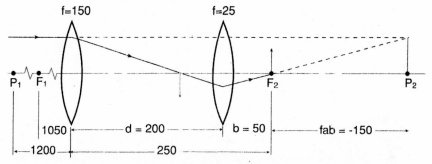

Figure 1.14 A widely separated system of two positive components with the second component acting as an erecting relay lens. This system forms a real, erect image and has a negative focal length. Note that the second principal point is to the right of the second focal point. The combination has a focal length equal to the focal length of the first lens times the magnification of the relay lens.

mm. Thus $f_{ab} = -150$. The sum of the space d and back focus B must be 250 mm. Thus $d = 250 - B$.

Substituting for d,
Eq. 1.31

$$f_a = \frac{(250 - B)(-150)}{-150 - B}$$

Eq. 1.32

$$f_b = \frac{-(250 - B)B}{-150 - 250}$$

Obviously we are free to select the value of B (or d). Arbitrarily setting $B = 50$, we get

$$f_a = \frac{(250 - 50)50}{-150 - 50} = +150.0 \qquad \phi_a = +0.00666...$$

$$f_b = \frac{-(250 - 50)50}{-150 - 250} = +25.0 \qquad \phi_b = +0.040$$

Additional study project: Calculate the component powers for several additional values of B. Plot ϕ_a, ϕ_b, and $(|\phi_a| + |\phi_b|)$ against B to find the minimum power required to do the job.

Sample calculation

Find the Gauss points for the system in the previous calculation. See Fig. 1.14.

Since we know the focal length is -150 mm and that B is 50, F_2 must be 50 mm to the right of component b. P_2 is 150 mm to the *right* of F_2 because the system focal length is negative; therefore, P_2 is 200 mm to the right of the rear component.

Using Eq. (1.30) we find $FF = -(-150)(25-200)/25 = -1050$, which indicates that F_1 is 1050 mm to the left of component a. This puts P_1 at $(-1050-150) = -1200$, or 1200 mm to the left of the first lens.

Sample calculation

We need a magnification of $+2\times$ in a distance of 0.9 m, using two components which are evenly spaced between object and image. Determine the necessary component powers. See Fig. 1.15A.

By the "evenly spaced" requirement we mean that s, d, and s' are all the same size. Then, by our sign convention, $s = -300$, $d = +300$, and $s' = +300$.

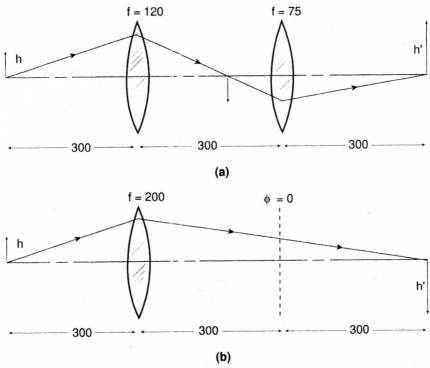

Figure 1.15 Two solutions to the problem of producing a two times magnification in an object-to-image distance of 900 mm. One solution has a positive magnification (and an erect image) and the other has a negative magnification (and an inverted image). It is common to find two different systems which produce images of the same size in the same place, one producing an erect image, the other an inverted image, if you use plus and minus magnifications in the equations.

By Eq. (1.33):

$$\phi_a = \frac{2(-300) - 2 \cdot 300 - 300}{2(-300)300} = +0.00833...$$

$$f_a = \frac{1}{\phi_a} = 120.0$$

By Eq. (1.34):

$$\phi_b = \frac{300 - 2(-300) + 300}{300 \cdot 300} = +0.0133...$$

$$f_b = \frac{1}{\phi_b} = 75.0$$

However, if we use a magnification of $-2\times$, so that the image is inverted, we get the following. See Fig. 1.15B.

By Eq. (1.33):

$$\phi_a = \frac{-2(-300)-(-2)300 - 300}{(-2)(-300)300} = +0.005$$

$$f_a = \frac{1}{\phi_a} = 200.0$$

By Eq. (1.34):

$$\phi_b = \frac{300-(-2)(-300) + 300}{300\cdot300} = 0.00$$

$$f_b = \frac{1}{\phi_b} = \infty$$

This indicates that if image orientation were not a concern, this particular task could be handled with just one lens rather than the two which are needed when the image is required to be erect. This illustrates the importance of considering both erect and inverted imagery if one is not preferred over the other.

Sample calculation

Lay out a Cassegrain mirror system with a focal length of 100, a mirror separation of 25, and an image distance of 30. Use Eqs. (1.31) and (1.32) with $f_{ab} = +100$, $d = 25$, and $B = 30$. See Fig. 1.16.

Note that when we *raytrace* the Cassegrain system, both the spacing and the index between the mirrors are considered to be negative (because the secondary mirror is to the left of the primary, and because light is traveling from right to left). The *equivalent air distance* is the spacing divided by the index, so, as explained in Sec. 2.13, we use $-25/-1.0 = +25$ for d in Eqs. (1.31) and (1.32).

By Eq. (1.31):

$$f_a = \frac{25\cdot100}{100 - 30} = 35.714$$

By Eq. (1.32):

$$f_b = \frac{-25\cdot30}{100-30-25} = -16.66...$$

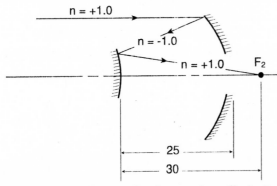

Figure 1.16 The Cassegrain mirror system, illustrating the convention of using a positive index of refraction for light traveling left to right, and a negative index when the light travels right to left (as after reflecting from the primary mirror).

Remembering that a mirror radius is twice its focal length, and that concave mirrors have positive (convergent) focal lengths, the primary mirror has a focal length of $+35.714$ and a concave radius of -71.428; the secondary mirror has a focal length of -16.667 and a convex radius of -33.333. Note that the raytrace sign of the radius is determined by the location of its center of curvature, not by whether the mirror is concave or convex.

1.11   The Scheimpflug Condition

To this point we have assumed that the object is defined by a plane surface which is normal to the optical axis. However, if the object plane is tilted with respect to the vertical, then the image plane is also tilted. The Scheimpflug condition is illustrated in Fig. 1.17A, which shows the tilted object and image planes intersecting at the plane of the lens. Or, stated more precisely for a thick lens, the extended object and image planes intersect their respective principal planes at the same height.

For small tilt angles in the paraxial region, it is apparent from Fig. 1.17A that the object and image tilt are related by

$$\Theta' = \Theta\, \frac{s'}{s} = m\Theta \tag{1.38a}$$

For finite (real) angles

$$\tan \Theta' = \frac{s'}{s} \tan \Theta = m \tan \Theta \tag{1.38b}$$

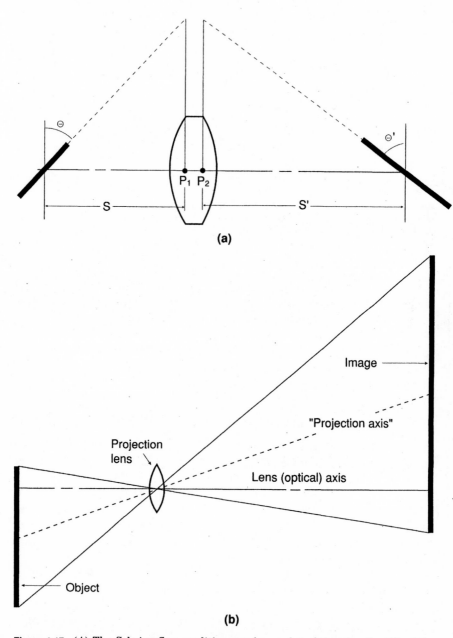

(a)

(b)

Figure 1.17 (*A*) The Scheimpflug condition can be used to determine the tilt of the image surface when the object surface is tilted away from the normal to the optical axis. The magnification under these conditions will vary across the field, producing "keystone" distortion. As diagramed here, the magnification of the top of the object is larger than that of the bottom. (Compare the ratio of image distance to object distance for the rays from the top and bottom of the object.) (*B*) Keystoning can be avoided if the object and image planes are parallel. The figure shows how the "projection axis" can be tilted upward without producing keystone distortion.

Note that in general a tilted object or image plane will cause what is called *keystone distortion*, because the magnification varies across the field. This results from the variation of object and image distances from top to bottom of the field. This distortion is often seen in overhead projectors when the top mirror is tilted to raise the image projected on the screen. This is equivalent to tilting the screen. As shown in Fig. 1.17B, keystone distortion can be prevented by keeping the plane of the object effectively parallel to the plane of the image. In a projector this means that the field of view of the projection lens must be increased on one side of the axis by the amount that the beam is tilted above the horizontal.

1.12   Reflectors, Prisms, Mirrors, etc.

Our initial assumption in Sec. 1.1 was of axial symmetry about the optical axis. The axis may be *folded* by reflection from a plane surface without losing the benefits of axial symmetry. This is often done to fit an optical system into a prescribed space or to get around an obstruction. It is also frequently done to change the orientation of the image, e.g., to turn it top to bottom and/or reverse it left to right.

The folding of the axis is easily understood through Snell's law of reflection, $I' = -I$, where I is the angle between the original axis ray and the normal to the reflecting surface, and I' is the corresponding angle after reflection. An example of a reflecting surface "folding" a system is shown in Fig. 1.18. Often a scale drawing of the entire optical system is made on tracing paper, and the paper is then folded to simulate the reflections.

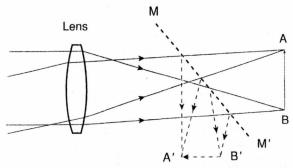

Figure 1.18 A mirror can be thought of as "folding" the optical system, just as if the paper were folded along the line indicating the mirror. The lens image at AB is the object for the mirror MM', and the mirror forms an image at $A'B'$, which is on the other side of the mirror and an equal distance behind it.

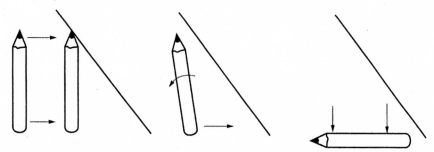

Figure 1.19 The orientation of the image formed by a mirror system can easily be determined by simply "bouncing" a pencil (or any object with one well-defined end) off the mirror as indicated here. In the figure the point of the pencil strikes the mirror first and bounces off; the blunt end strikes later. The orientation of the pencil indicates the orientation of the image. This can be done for a sequence of mirrors to determine the final image orientation. It should be done for both meridians.

The orientation of the image can be determined easily by use of the simple "bouncing pencil" technique as illustrated in Fig. 1.19. This technique can be applied to a sequence of reflections to determine the final orientation of the image; it should be applied to both meridians as shown in Fig. 1.20.

In most instances a reflecting system can be executed with either prisms or first-surface mirrors. In a mirror system there is a small reflection loss at each surface, but the system is light in weight and does not require a good transmissive material. A prism system usually reflects by *total internal reflection* (TIR), which is 100 percent efficient and which occurs when the angle of incidence exceeds the *critical angle* $I_c = \arcsin(n'/n)$. This occurs only when light in a higher-index material is incident on a surface with a lower index on the other side (e.g., from glass into air). For $n' = 1.0$ and $n = 1.5$, $I_c = 42°$; for $n = 1.7$, $I_c = 36°$. The drawback to a prism system is its weight and the necessity for a high-quality optical material.

Most prism systems are the equivalent of a thick folded glass block, as indicated in Fig. 1.21. A glass block (or plane-parallel plate) shifts the image to the right by $(n-1)t/n$, or about one-third of its thickness. It also introduces overcorrected spherical and chromatic aberrations.

As illustrations of image displacement and reorientation, two common *erecting systems* are shown in Fig. 1.22. Note that in the Porro 1 system, the first prism inverts the image top to bottom and the second prism reverses it left to right. Both systems have four-mirror equivalents.

Three *inversion prisms* are shown in Fig. 1.23. These invert the image in one meridian but not in the other. An inversion prism has

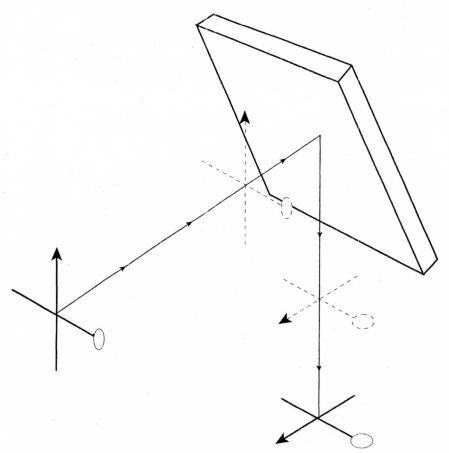

Figure 1.20 The two-meridian orientation change produced by the reflection from a single mirror.

the property that if it is rotated about the optical axis, the image is rotated at twice the angular rate. They are often used as *derotators* in applications such as panoramic telescopes, where the 360° scan rotates the image. Note that the addition of a "roof" surface will convert an inversion system to an erecting system, and vice versa.

A pair of reflecting surfaces can be used as a *constant deviation* system, as shown in Fig. 1.24. The angle of deviation in the plane of the paper is twice the angle between the reflectors, and it does not change as the angle of incidence is changed. In three dimensions, three mutually orthogonal reflectors, arranged like the corner of a cube, form a *retrodirector* and reflect light back parallel to the direction at which it entered the device.

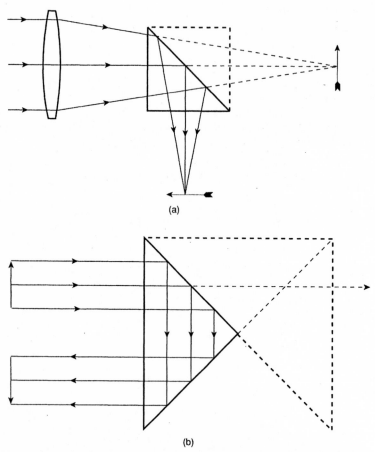

(a)

(b)

Figure 1.21 "Unfolding" a prism system produces what is called a "tunnel diagram," which is useful in determining the clearance available for the passage of rays through the prisms. These sketches also illustrate the equivalence of the prism to a thick block, or plane-parallel plate, of glass.

1.13   Collected Equations

Imaging equations

Figure 1.3
Eq. (1.1)

$$x' = \frac{-f^2}{x}$$

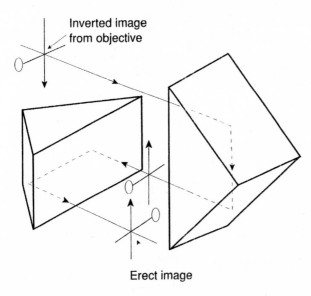

Erect image

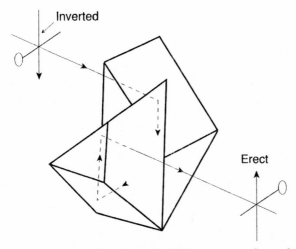

Figure 1.22 Two prism systems which are commonly used to erect the inverted image produced by an ordinary "keplerian" telescope. As can be seen, these prisms invert the image in both meridians. The upper system is a Porro 1 type which is commonly found in ordinary binoculars. The lower is a Porro 2; it is a bit more compact and a bit more expensive. Note that the same image orientation could be produced with four first-surface mirrors replacing the reflecting faces of the prisms.

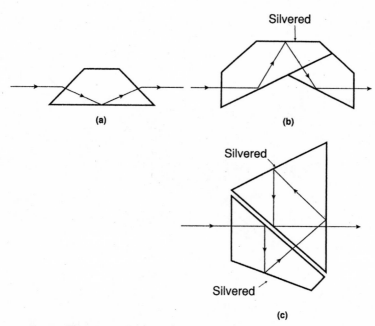

Figure 1.23 The prisms shown here are called inversion prisms. They invert the image in one meridian but not the other. All inversion prisms share the property that if rotated about the axis, they rotate the image at twice the prism's rate of rotation.

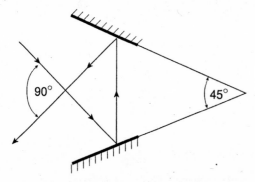

Figure 1.24 Two reflecting surfaces can produce a constant deviation system, which deviates the ray through the same angle regardless of the direction at which the ray enters the system. Here the mirrors, set at 45° to each other, produce a deviation of 90°. The deviation angle is twice the angle between the reflectors.

Eq. (1.2)

$$h' = \frac{hf}{x} = \frac{-hx'}{f}$$

Eq. (1.3)

$$m = \frac{h'}{h} = \frac{f}{x} = \frac{-x'}{f}$$

Eq. (1.4)

$$\frac{1}{s'} = \frac{1}{f} + \frac{1}{s}$$

Eq. (1.5)

$$h' = \frac{hs'}{s}$$

Eq. (1.6)

$$m = \frac{h'}{h} = \frac{s'}{s}$$

$$s' = \frac{sf}{s + f}$$

$$s' = f(1 - m)$$

$$s = \frac{f(1 - m)}{m}$$

$$f = \frac{ss'}{s - s'}$$

Figure 1.4
Eq. (1.7)

$$h' = fu_p$$

Eq. (1.8)

$$H' = f \tan U_p$$

Figure 1.5
Eq. (1.9)

$$M = \frac{s_2' - s_1'}{s_2 - s_1}$$

Eq. (1.10)

$$M = \frac{s_1'}{s_1} \cdot \frac{s_2'}{s_2} = m_1 m_2$$

Eq. (1.11)

$$M = m^2$$

Paraxial raytracing
Figure 1.8
Eq. (1.12)

$$l = \frac{-y}{u}$$

Eq. (1.13)

$$l' = \frac{-y}{u'}$$

Eq. (1.14a)

$$n'u' = nu - \frac{y(n' - n)}{r}$$

Eq. (1.14b)

$$n'u' = nu - y(n' - n)c$$

Eq. (1.15)

$$y_2 = y_1 + tu_1'$$

Eq. (1.16a)

$$m = \frac{h'}{h} = \frac{n_1 u_1}{n_k' u_k'}$$

Eq. (1.16b)

$$m = \frac{h'}{h} = \frac{u_1}{u_k'}$$

Figure 1.9
Eq. (1.17)

$$f = efl = \frac{-y_1}{u_k{}'}$$

Eq. (1.18)

$$bfl = \frac{-y_k}{u_k{}'}$$

Thin lens raytracing
Figure 1.10
Eq. (1.19)

$$u' = u - y\phi$$

Eq. (1.20)

$$y_{j+1} = y_j + du_j{}'$$

The invariant
Eq. (1.21)

$$\text{INV} = n(y_p u - y u_p) = n'(y_p u' - y u_p{}')$$

Eq. (1.22)

$$\text{INV} = hnu = h'n'u'$$

Scaling and combining rays
Eq. (1.23)

$$u_3 = Au_1 + Bu_2$$

Eq. (1.24)

$$y_3 = Ay_1 + By_2$$

Eq. (1.25)

$$A = \frac{y_3 u_1 - u_3 y_1}{u_1 y_2 - y_1 u_2}$$

Eq. (1.26)

$$B = \frac{u_3 y_2 - y_3 u_2}{u_1 y_2 - y_1 u_2}$$

Combination of two components

Figure 1.11

Eq. (1.27)

$$\phi_{ab} = \phi_a + \phi_b - d\phi_a\phi_b$$

Eq. (1.28)

$$f_{ab} = \frac{f_a f_b}{f_a + f_b - d}$$

Eq. (1.29)

$$B = \frac{f_{ab}(f_a - d)}{f_a}$$

Eq. (1.30)

$$FF = \frac{-f_{ab}(f_b - d)}{f_b}$$

Eq. (1.31)

$$f_a = \frac{df_{ab}}{f_{ab} - B} = \frac{1}{\phi_a}$$

Eq. (1.32)

$$f_b = \frac{-dB}{f_{ab} - B - d} = \frac{1}{\phi_b}$$

Figure 1.12

Eq. (1.33)

$$\phi_a = \frac{ms - md - s'}{msd}$$

Eq. (1.34)

$$\phi_b = \frac{d - ms + s'}{ds'}$$

Eq. (1.35)

$$0 = d^2 - dT + T(f_a + f_b) + \frac{(m-1)^2 f_a f_b}{m}$$

Eq. (1.36)

$$s = \frac{(m-1)d + T}{(m-1) - md\phi_a}$$

Eq. (1.37)

$$s' = T + s - d$$

Scheimpflug condition

Figure 1.17A

Eq. (1.38a)

$$\Theta' = \Theta \frac{s'}{s} = m\Theta$$

Eq. (1.38b)

$$\tan \Theta' = \frac{s'}{s} \tan \Theta = m \tan \Theta$$

2

The Basic
Optical Systems

2.1 Introduction

The optics used for most applications are founded on what might be called the basic "standard" optical systems. And, in fact, almost all optical systems are modifications or combinations of these "standard" or basic systems. The principles of these systems are well understood, and their primary characteristics and limitations can be expressed by simple mathematical relationships. This chapter is intended as an exposition of the concepts and principles of these basic systems. The reader can utilize these systems as building blocks to synthesize a customized solution to the requirements of the application at hand.

2.2 Stops and Pupils

We begin with a discussion of an often neglected but vital aspect of optical systems, the concept of the *aperture stop*. In every optical system there is some feature, usually a diameter, which limits the size or diameter of the beam of rays which can pass through the system. If the system consists of just a single simple lens, the aperture stop is just the clear aperture of the lens. In a typical camera lens the iris diaphragm is the aperture stop. In a telescope the clear aperture of the objective lens (or that of the primary mirror) is ordinarily the aperture stop. Often the aperture stop is referred to simply as "the stop."

Figure 2.1 illustrates several examples of aperture stops. Note that Fig. 2.1C and D shows exactly the same system, except that the objective lens diameter is larger in Fig. 2.1D. This shifts the beam-limiting

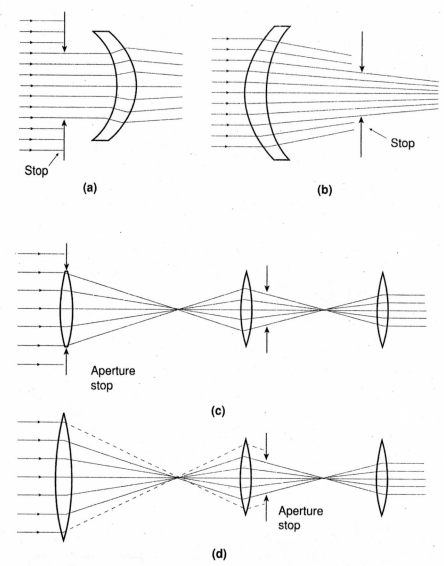

Figure 2.1 The aperture stop: (A) The aperture stop placed before a single element. (B) The aperture stop placed behind the lens. (C) An erecting telescope with the aperture stop located at the objective lens. (D) The same optics as in C, but the objective aperture is larger so that the aperture stop is now at the internal aperture.

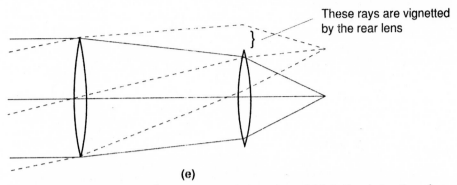

These rays are vignetted by the rear lens

(e)

Figure 2.1 (*Continued*) (*E*) An illustration of vignetting, which is the obstruction of a portion of an oblique beam by the aperture(s) of the components.

diameter from the objective lens to the internal diaphragm, which becomes the stop. The stop can be determined by tracing a fan or cone of rays from the axial object point and determining which feature most limits the size of the ray bundle.

An effect often produced by the ordinary apertures of optical components is called *vignetting.* This is the partial obstruction of an oblique beam of light by the diameters of the lenses. Figure 2.1*E* shows a simple two-component system with an axial beam indicated by the solid-line rays and an oblique beam shown by the dashed lines. Notice that the full axial beam passes through both components, but the upper rays of the oblique beam are vignetted (i.e., blocked) at the rear lens. Vignetting usually occurs at the first and/or last elements of an assembly. Vignetting reduces the image illumination toward the edge of the field (as does the cosine fourth effect described in Sec. 3.6).

Any image of the aperture stop is called a *pupil.* This image is formed by the optical elements of the system. Figure 2.2 shows the imagery of the stop for the systems of Fig. 2.1. We trace a ray from the center of the stop; wherever the ray (or its extension) crosses the axis is the location of a pupil. There are two especially significant pupils: the *entrance pupil* and the *exit pupil.* These are the images of the aperture stop that one would see looking into the system from the object and from the image, respectively.

Note that in Fig. 2.2*C* the internal diaphragm is not located at the internal pupil; this has been done to more clearly illustrate the entrance pupil in Fig. 2.2*D*. In most real systems of this type the internal diaphragm and the internal pupil are deliberately made coincident, and the objective lens is both the aperture stop and the entrance pupil.

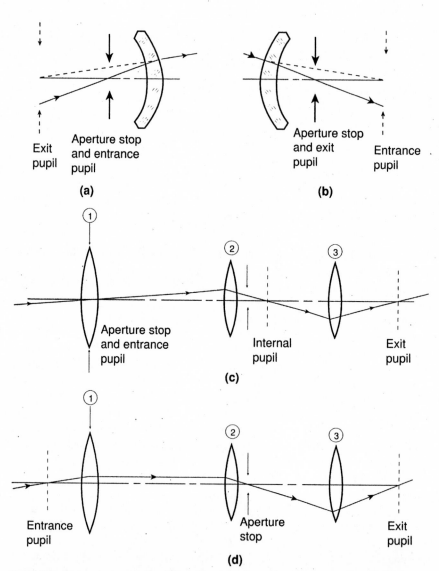

Figure 2.2 The entrance and exit pupils of the systems of Fig. 2.1. (A) The entrance pupil and the aperture stop are the same. The exit pupil is the virtual image of the stop, which one would see looking into the lens from the right. (B) Here, the aperture stop and the exit pupil are the same, and the entrance pupil is the virtual image of the stop as seen from object space. (C) Every image of the stop is a pupil, and here there is an internal pupil as well as an exit pupil. (D) With the internal stop acting as the aperture stop, the entrance pupil is its image and is located to the left of the objective in object space.

Sample calculation

Locate the entrance pupil for Fig. 2.2B if the lens focal length is 100 mm and the stop is 20 mm to its right. To calculate this, we must reverse the system so that the light travels from left to right; this puts the stop 20 mm to the *left,* and the object distance $s = -20$.

Using Eq. (1.4):

$$\frac{1}{s'} = \frac{1}{f} + \frac{1}{s} = \frac{1}{100} + \frac{1}{-20} = -0.04$$

$$s' = \frac{1}{-0.04} = -25 \text{ mm}$$

With the lens in this *reversed* position, the image of the stop is 25 mm to the left, indicating that, in the original orientation (as shown in Fig. 2.2B), the stop image, and the entrance pupil, is 25 mm to the right of the lens. This simple example illustrates the importance of observing the convention (Sec. 1.1, no. 4) under which these equations are derived: Light rays are assumed to travel from left to right.

Sample calculation

Locate the internal pupil and the exit pupil for the telescope shown in Fig. 2.2C, assuming that the lens powers are $+0.25$, $+1.0$, and $+1.0$, and that the spacings are 6.0 and 3.0 in. The aperture stop is at the first lens and it is 1.0 in in diameter. How large are the pupils? Here we can utilize Eqs. (1.19) and (1.20) (Sec. 1.7) to trace a chief ray from the center of the aperture stop. Assuming a ray slope of $u_1' = u_2 = 0.1$ (any value would do), Eq. (1.20) gives the ray height at lens 2 as

$$y_2 = y_1 + du_2 = 0.0 + 6.0 \times 0.1 = 0.6$$

The ray slope after refraction by lens 2 is

$$u_2' = u_2 - y_2\phi_2 = 0.1 - 0.6 \times 1.0 = -0.5$$

The internal pupil is located at

$$l_2' = \frac{-y_2}{u_2'} = \frac{-0.6}{-0.5} = +1.2$$

to the right of lens 2. The magnification is found from Eq. (1.16):

$$m = \frac{h'}{h} = \frac{u_2}{u_2'} = \frac{0.1}{-0.5} = -0.2$$

Since the aperture stop diameter is 1.0 in, its image has a (pupil) diameter of

$$h' = mh = -0.2 \times 1.0 = -0.2 \text{ in}$$

Continuing the raytrace, Eq. (1.20) gives the ray height at lens 3 as

$$y_3 = y_2 + du_2' = 0.6 + 3(-0.5) = 0.6 - 1.5 = -0.9$$

and Eq. (1.19) gives the ray slope after lens 3 as

$$u_3' = u_3 - y_3\phi_3 = -0.5 - (-0.9) \times 1.0 = +0.4$$

Thus the exit pupil has a location and size given by

$$l_3' = \frac{-y_3}{u_3'} = \frac{-(-0.9)}{0.4} = 2.25 \text{ (eye relief)}$$

$$m = \frac{h'}{h} = \frac{u_2}{u_3'} = \frac{0.1}{0.4} = 0.25$$

$$h' = mh = 0.25 \times 1.0 = 0.25 \text{ (exit pupil diameter)}$$

The aperture stop is of practical significance on two counts. It is obvious that if light is to pass through the system it must pass through the stop; since the pupils are images of the stop, all the light must also pass through the pupils. Thus the stop and the pupils determine the radiometric and photometric characteristics of the system. Light must enter the system through the entrance pupil and must leave through the exit pupil. The stop also selects which light rays out of all the possible rays from the object are allowed to reach the image. A properly located aperture stop will pass those rays which are least aberrated and which will form the best image. The stop location is an important factor in determining the quality of the image in the outer portions of the field. The stop position also determines the lens diameter necessary to pass the oblique rays through the system.

The *relative aperture,* or *f-number,* of a lens is a way of describing its illuminating capabilities. The *f*-number of a lens (often called its *speed*) is simply its effective focal length divided by its entrance pupil diameter. If the object is at infinity, the *f*-number defines the angular size of the illuminating cone at the image and thus, as described in Chap. 3, the image illumination (and also, as described in Chap. 4, the resolution capability of the system).

The *numerical aperture,* or NA, is a more general way of defining the same characteristics and is not limited to systems with infinitely distant objects. It is defined as

$$NA = n \sin u$$

where n is the index in which the image is immersed (with the image in air, $n = 1.0$) and u is the slope of the axial marginal ray. For systems with infinitely distant objects, *f-number* and NA are related by

$$f\text{-number} = \frac{1}{2NA}$$

$$NA = \frac{1}{2f\text{-number}}$$

The *working f-number* is sometimes used to describe systems with finite object distances; it is the image distance divided by the pupil diameter and can be found from working f-number $= 1/(2NA)$. Note also that object side NA and image side NA are related by the magnification, as indicated by Eq. (1.16).

Sample calculation

A simple lens with a 10-in focal length and a 1=in diameter has a speed of $f/10$ and a numerical aperture of NA $= 0.5/10 = 0.05$ if used with an object at infinity. But at one-to-one magnification ($m = -1.0$), $s = -20$ in and $s' = +20$ in. Then the numerical aperture NA $= 0.5/20 = 0.025$ and the working f-number is $f/20$.

In a visual optical system the eye must look through the exit pupil to see the image. The situation is analogous to looking through a hole in a sheet of opaque material. In order to see the full field of view, the eye must be located at the exit pupil, as shown in Fig. 2.3. If the eye is not located at the pupil, only a part of the field may be visible. The distance from the final optical surface to the exit pupil represents the clearance distance to the eye. This distance is called the *eye relief*. The eye relief must be 9 or 10 mm just to clear the eyelashes; it must be at least 19 or 20 mm to permit spectacle wearers to place their eyes at the pupil; it must be much longer for an optical system which is subject to sudden motion. For example, a riflescope should have at least 50-mm eye relief for a low-powered (0.22 caliber) rifle, and needs 100- to 150-mm eye relief for a high-powered gun with a large recoil.

Sample calculation

In a previous sample calculation, the eye relief of a 4× telescope was found to be 2.25 in and the exit pupil diameter was 0.25 in. If lens 3

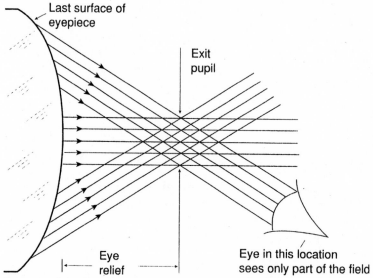

Figure 2.3 In a visual instrument the clearance between the optics and the exit pupil is called the eye relief. From the diagram it is apparent that if the eye is not at the exit pupil, only a portion of the field will be visible. To see the full field of view, the eye must be positioned at the exit pupil.

has a clear aperture of 1.25 in, what is the largest apparent field possible without any vignetting (the obstruction of an oblique beam)?

If we deduct the exit pupil diameter from the lens aperture and divide by the eye relief, we find a paraxial apparent field angle of

$$A' = \frac{1.25 - 0.25}{2.25} = 0.444...$$

The real (object) field of a 4× telescope can be found from Eqs. (2.1) or (2.2) below. Thus we get a paraxial real field angle of

$$A = \frac{A'}{MP} = \frac{0.444}{4} = 0.111...$$

If we consider finite angles (as opposed to the infinitesimal paraxial slopes), we must use the arctangent of the half field angles [per Eq. (2.2)] to get

$$A' = 2 \arctan\left(\frac{0.444}{2}\right) = 2 \times 12.52 = 25.05°$$

$$A = 2 \arctan\left(\frac{0.111}{2}\right) = 2 \times 3.18 = 6.36°$$

A *telecentric system* is one in which the exit pupil or the entrance pupil (or both) is located at infinity. Thus the telecentric aperture stop must be positioned at the focal point of that part of the system which forms the pupil. Referring ahead to Fig. 2.5A, if one were to locate the aperture stop of the system at the internal focal point (instead of at the objective lens as shown), the system would be doubly telecentric, with both entrance and exit pupils at infinity. The solid-line axial rays in the figure would then be principal rays and would be parallel to the axis in both object and image space. The advantage of a telecentric system is that, if it is used out of focus, the image may be blurred, but it doesn't change size (because the "principal" rays are parallel to the axis). This is a valuable characteristic for measuring systems and for microlithographic (computer chip fabricating) optical systems.

The *field stop* limits the size of the object that the system can image. The field stop is almost always located at an internal image plane. In a camera the film gate is the field stop. In most visual instruments (telescopes and microscopes) there is a diaphragm which sharply defines and limits the field of view; it is located at an internal image plane. Such a field stop is shown in Figs. 2.5C and 2.7.

The entrance and exit *windows* of a system are the images of the field stop in object and image space, respectively. Since they are usually coincident with the object and image, their significance is ordinarily negligible, and these terms are only rarely encountered.

2.3 Afocal Systems: General

An afocal system is one which images an infinitely distant object at infinity. It is called afocal because it has no focal length (or an infinite one; either concept can be a nuisance). Telescopes, beam expanders, power and field changers, as well as telephoto and wide-angle attachments, are examples of afocal systems.

An afocal system can form an (infinitely distant) image which subtends an angle which is larger or smaller than the angle subtended by the object. The ratio of the angle subtended by the image to that subtended by the object is the *angular magnification,* or *magnifying power,* MP, of the afocal system. If the image is enlarged, as in a telescope, $|MP| > 1.0$; if the image is smaller than the object, $|MP| < 1.0$. When the image appears erect, MP is positive; a negative MP indicates an inverted image.

The magnification MP, the diameters of the entrance pupil P and exit pupil P', and the angular size of the real (object side) field A and the apparent (image side) field A' are all related by the following equation:

$$MP = \frac{A'}{A} = \frac{P}{P'} \tag{2.1}$$

Equation (2.2) gives this relationship for finite (as opposed to infinitesimal/paraxial) field angles:

$$MP = \frac{\tan(A'/2)}{\tan(A/2)} = \frac{P}{P'} \tag{2.2}$$

All afocal systems must obey these relationships. For example, in a typical 6× binocular, the exit pupil diameter must be one-sixth of the entrance pupil diameter, and the apparent (image) field must be six times the real (object) field. A 6×30 binocular which has a 30-mm-diameter objective aperture must have a 5-mm exit pupil. If the eyepiece covers a 48° field of view, then the real field at the object must be about 48/6 = 8° [or, more exactly, using Eq. (2.2), 8.5°].

Figure 2.4 shows a schematic afocal system. Because the object and image are at infinity, the rays from an axial object point and those going to the image point are all parallel to the axis. Although we define an afocal system as one with both object and image at an infinite distance, an afocal system is also capable of imagery at finite distances. For example, the entrance and exit pupils are conjugates of each other, and an examination of Fig. 2.4 indicates that their sizes are determined by the rays drawn as heavy lines in the figure. Since the rays (or lines) are parallel, it is apparent that regardless of where the pupils are located, their size stays the same. The exit pupil may be regarded as an image of the entrance pupil. The (linear) magnification of the pupils is then equal to P'/P, which, per Eq. (2.1), is the reciprocal of the angular magnification MP. Thus:

$$m = \frac{P'}{P} = \frac{1}{MP} \tag{2.3}$$

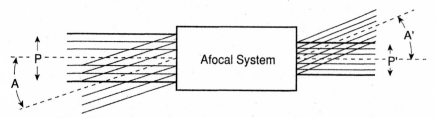

Figure 2.4 Schematic of an afocal system, showing the entrance and exit pupil diameters P and P' and the angular half fields A and A', which are the "real" and "apparent" field, respectively.

An afocal system which is used for finite conjugate imagery is thus a constant magnification system because, regardless of where the object is positioned, the linear magnification is always the same.

Afocal systems are ordinarily considered as made up of two subsystems: the objective and the eyepiece, the objective being the portion nearest the object and the eyepiece the part nearest the image. If F_o is the effective focal length of the objective part and F_e is the *efl* of the eyepiece part, then the angular magnification is given by

$$MP = -\left(\frac{F_o}{F_e}\right) \qquad (2.4)$$

Although the objective and eyepiece are ordinarily obviously independent parts, note that the division between objective and eyepiece may be completely arbitrary. The dividing line may be placed anywhere after the first powered surface and before the last; this fact is occasionally a convenience in layout considerations. And, of course, when an afocal device is reversed, as in looking through a binocular backward, the objective and eyepiece are interchanged, and the magnifying power becomes 1/MP.

2.4   Telescopes and Beam Expanders

Kepler or astronomical telescope.   As shown in Fig. 2.5, in the keplerian telescope both the objective and eyepiece have positive focal lengths. Thus, per Eq. (2.4), the magnification is negative, indicating that the image is inverted; it is upside down and reversed left to right. For terrestrial use, as in binoculars, for example, where an inverted image is quite undesirable, the image is often erected by either an erector lens or a prism system, as discussed in Sec. 1.12. If the lenses are "thin," the length of the system is given by

$$L = F_o + F_e \qquad (2.5)$$

The exit pupil is real and accessible, and the eye relief (again, for thin lenses) is

$$R = \frac{(MP - 1)F_e}{MP} \qquad (2.6)$$

Sample calculations

Lay out a 10-in-long, $4\times$ Kepler telescope. Find the eye relief. Equation (2.4) defines the magnification, Eq. (2.5) the length, and Eq.

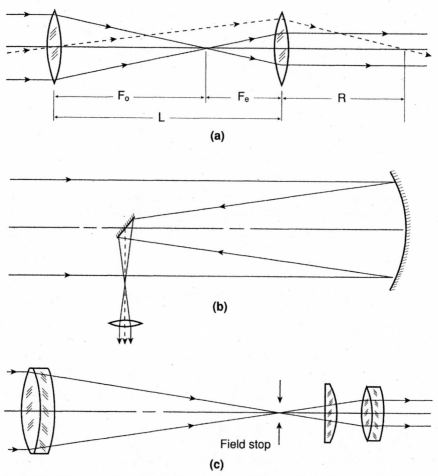

Figure 2.5 Kepler-type telescopes produce an inverted image. (A) A schematic sketch of a simple two- (positive) component telescope with the focal lengths F_o and F_e' overall length L, and eye relief R. (B) The newtonian-type telescope has a reflecting objective and a 45° mirror to locate the image out of the incoming beam of light. (C) An ordinary telescope with an achromatic objective lens and a three-element eyepiece. The field stop at the internal focus determines the angular field of view. With a Porro prism to erect the image (as shown in Fig. 1.22), this is the optical system commonly found in ordinary binoculars.

(2.6) the eye relief. We can solve Eqs. (2.4) and (2.5) simultaneously to get the component powers as follows:

$$F_o = \frac{MP \times L}{MP - 1} \qquad (2.7)$$

$$F_e = \frac{L}{1 - MP} \qquad (2.8)$$

Note that since the Kepler is an inverting telescope we must use MP $= (-4\times)$ as the power, and we get

$$F_o = \frac{-4 \times 10}{-4-1} = \frac{-40}{-5} = +8 \text{ in}$$

$$F_e = \frac{10}{1 + 4} = \frac{10}{5} = +2 \text{ in}$$

Equation (2.6) gives the eye relief as

$$R = \frac{(-4 - 1)2}{-4} = \frac{-10}{-4} = 2.5 \text{ in}$$

Galilean or Dutch telescope. The galilean telescope (Fig. 2.6) has a positive objective and a simple (i.e., not compound) negative eyepiece. The angular magnification is positive and the image is erect. The field of view of this telescope tends to be small and is limited by the speed (f-number) of the objective. As a result its telescopic use is limited to low-power field glasses (of $3\times$ to $4\times$ power) or opera glasses ($1.5\times$ to $2.0\times$); the real field of a high-powered galilean telescope can be so small as to be useless. Another limitation of the galilean is that, since there is no internal focal point, a crosshair or reticle cannot be used; the Kepler or terrestrial forms are used when a reticle is needed.

As with the keplerian scope, the power and length are given by Eqs. (2.4) and (2.5). Note that the eye relief as indicated by Eq. (2.6) is the

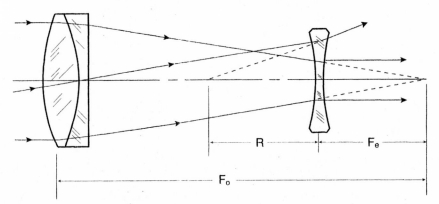

Figure 2.6 The galilean telescope, commonly used for field or opera glasses, produces an erect image. As indicated in the sketch, if the objective were the aperture stop, the exit pupil would lie inside the system and be inaccessible to the eye. For this reason the pupil of the eye acts as the aperture stop, and the objective lens diameter determines the size of field of view (which is usually small).

image of the objective lens; it assumes that the objective lens is the aperture stop. For the galilean telescope this equation yields a negative distance, indicating that the image of the objective is virtual, and located inside the telescope as shown in Fig. 2.6. The eye obviously cannot be placed here; the closest it can approach is adjacent to the eyelens. Thus the objective lens cannot be the aperture stop; the pupil of the eye itself becomes both the aperture stop and the exit pupil. The largest possible apparent field is the angle subtended by the objective diameter from the eyelens; the real field is this angle divided by the magnifying power MP. Apparent and real fields for the galilean are

$$A' = \frac{F_o}{L(f/\#)} \tag{2.9}$$

$$A = \frac{F_o}{L(f/\#)\text{MP}} = \frac{-F_e}{L(f/\#)} \tag{2.10}$$

where $(f/\#)$ is the f-number (focal length divided by diameter) of the objective lens.

Sample calculations

Lay out a 10-in-long $4\times$ galilean telescope. What is the maximum field if the objective diameter is 1.0 in? Using Eqs. (2.7) and (2.8) from the previous sample calculation with a power MP = $+4\times$, we get

$$F_o = \frac{4 \times 10}{4 - 1} = \frac{40}{3} = 13.333...$$

$$F_e = \frac{10}{1 - 4} = \frac{10}{-3} = -3.333...$$

At a diameter of 1 in, the objective lens f-number is $13.333/1 = 13.333$ (or $f/13.333$) and the maximum paraxial fields are given by Eqs. (2.9) and (2.10) as

$$A' = \frac{13.333}{10 \times 13.333} = 0.1$$

$$A = \frac{+3.333}{10 \times 13.333} = 0.025$$

or about 5.7° and 1.4°, respectively.

Terrestrial or lens-erecting telescope. The inverted image of the keplerian telescope can be erected by using a relay system as shown in Fig. 2.7. The magnification of the telescope can be determined from Eq. (2.4); one can consider the objective part of the scope to consist of the objective lens plus the erector, using Eq. (1.28) to determine its focal length; or one can consider the eyepiece part to consist of the eyelens plus the erector. Alternately, the magnification is given by

$$MP = -\left(\frac{F_o}{F_e}\right)\left(\frac{S'}{S}\right) \tag{2.11}$$

where S and S' are, respectively, the object and image distances for the erector component and, as shown in Fig. 2.7, the sign of S is negative per our usual sign convention.

Note that one can regard the focal length of the objective part to be the focal length of the objective lens multiplied by the magnification of the erector, or $F_o(S'/S)$. This yields a negative value for the focal length of the combination, despite the fact that it forms a real image. This concept of magnifying the focal length of a lens by the magnification of a relay lens is often a useful one.

The dashed ray passing through the center of the objective in Fig. 2.7 is a *principal* or *chief ray*. (This assumes that the objective is the aperture stop.) A pupil is located wherever the principal ray crosses the axis. In the terrestrial telescope we have the usual exit pupil and, in addition, an internal pupil. It is common, very beneficial, and usually essential to place an aperture at this location. Such an aperture, acting as a *glare stop*, serves to block the passage of any light which is reflected or scattered from the walls of the telescope. If this stray light is not blocked, the image contrast may be severely degraded. In general it is a good idea to place a stop at every pupil and at every image plane in order to minimize the effects of stray light.

Sample calculations

Lay out a 4× terrestrial telescope, 10 in long, with an objective focal length $F_o = 4$ in. We have $L-F_o = 10$ in-4 in $= 6$ in for the erector-eyelens combination; this includes the working distance (S in Fig. 2.7) and the space between the erector and eyelens. With an objective focal length of 4 in, the combined focal length of erector and eyelens must equal -1 in in order to yield MP $= +4×$. We can use Eqs. (1.31) and (1.32), letting the eyelens be component a and the erector be component b. Then B (S in Fig. 2.7) is the distance from the erector to the focus of the objective and d is the space; thus $B+d = 6$ in, or $d = 6-B$.

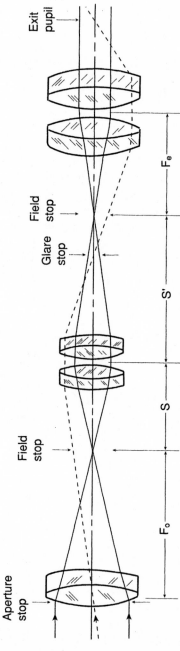

Figure 2.7 The terrestrial, or lens-erecting telescope. The image is relayed and erected by the central component. The aperture stop is usually at the objective lens, and a glare stop is placed at the internal pupil to block any stray light reflected from the walls of the telescope. There are two possible positions for the field stop (and a crosshair or reticle) in this system. The configuration shown here might be suitable for a riflescope.

From Eq. (1.31):

$$f_a = \frac{(6-B)(9-1)}{-1-B}$$

and Eq. (1.32):

$$f_b = \frac{-(6-B)B}{-1-B-(6-B)}$$

$$= \frac{6-B}{7}$$

It is apparent that we can choose any reasonable value for the working distance B. If we set $B = 1.5$ in, we get for the eyelens

$$f_a = \frac{(6-1.5)(-1)}{-1-1.5} = \frac{-4.5}{-2.5} = +1.8 \text{ in}$$

and

$$f_b = \frac{(6-1.5)1.5}{7} = +0.9643 \text{ for the erector}$$

We can use B as a free variable to control the eye relief if we wish. The initial choice of the objective lens focal length will affect the component powers. A few numerical trials can be used to determine how the choice of B and F_o affects the eye relief and the component powers. Note also that a field lens (as described in Sec. 2.7) can be used to control the eye relief of a telescope (or the exit pupil position for an optical system in general).

Beam expanders. Any afocal system can be utilized as a beam expander. Typically, a laser beam is directed into the "eyepiece end" of the device and the beam diameter is increased by a factor equal to the angular magnification $MP = -F_o/F_e$ of the afocal system. This not only expands the beam diameter but it also reduces the beam divergence angle by the same factor. Note that the divergence reduction applies either to the beam spread considered as a geometric field angle or to the beam spread due to diffraction. (See Sec. 4.2.) Either the keplerian or galilean form may be used; however, the galilean has certain advantageous features. Because it is comprised of a positive objective and a negative eyepiece, their aberrations have a tendency to cancel each other, whereas in the Kepler arrangement the aberrations of objective and eyepiece tend to add because both components are positive. For galilean beam expanders of low or moderate power a well-corrected system can be made of just two (properly shaped) simple elements. In addition, because it has no internal focal point, the galilean can be used to expand a high-powered laser beam without fear that atmospheric breakdown will occur at the focus. The Kepler does have the feature that a spatial filter (pinhole) can be placed at the internal focus to strip off any unwanted higher-order diffraction artifacts from the beam.

2.5   Afocal Attachments: Power and Field Changers

As shown in Fig. 2.8, an afocal device may be placed before another optical system to change its effective focal length. The resulting focal length for the combination F_c is simply the product of the afocal magnification MP and the focal length of the original system F_p. Thus

$$F_c = \text{MP} \cdot F_p \qquad (2.12)$$

From the figure, it can be seen that, depending on its orientation, the afocal attachment can produce a combined focal length F_c which is longer or shorter than the focal length of the prime lens F_p. When used as an attachment to a camera lens, the systems of Fig. 2.8 are usually referred to as telephoto or wide-angle attachments. Note that, with a galilean afocal camera attachment, the iris of the camera lens is the aperture stop of the system. Occasionally a binocular (which is a keplerian telescope) is used as a telephoto attachment. The real exit pupil of the binocular may cause problems because, for this arrangement to work well, the exit pupil of the binocular and the entrance pupil of the camera lens must be coincident. If they are not coincident, the field of view may be limited because of vignetting. With a binocular as the afocal, it is usually best to use a camera with a fast lens set at full aperture, so that the exit pupil of the binocular determines the *f*-number of the combination. Consider a 50-mm *f*/2 camera lens used with a 6×30 binocular. The combined focal length is 6×50 = 300 mm, and the 30 mm diameter of the binocular objective limits the speed of the combination to *f*/10.

An afocal is sometimes used as a "front end" for another system when a large collection aperture is required to "funnel" light into a smaller following system. One must not forget that the field angle is increased by the same magnification factor (MP) as that by which the beam diameter is reduced. This is another manifestation of the optical invariant.

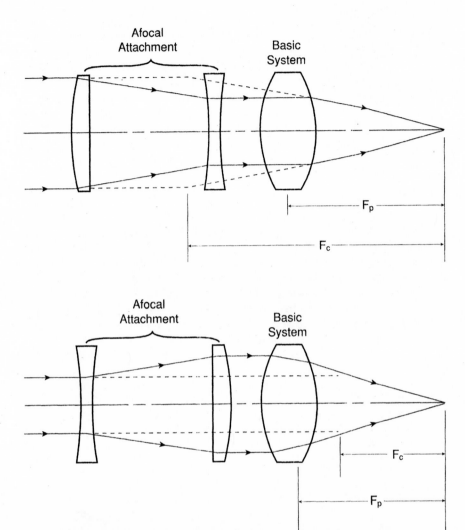

Figure 2.8 An afocal attachment can be placed in front of another system to change its focal length, power, or field. Shown here is a galilean afocal which, in the upper sketch, increases the focal length by a factor equal to its magnification and, when reversed, shortens the focal length by the same factor.

An afocal can be used before or after another afocal system. The resulting magnifying power is simply the product of the two magnifications. Thus the added afocal can be regarded as a power changer. It will also change the field of view by the inverse of its magnifying power. A 2× afocal attachment added to a 4× telescope with a 10° FOV (field of view) will produce an 8× scope with a 5° FOV. If the attachment is reversed, the result is a 2× scope with a 20° FOV. Thus the afocal is both a power changer and a field changer.

If there is a space in a system where the light is collimated, an afocal may be inserted within the system. "Collimated" means that the image is at infinity, so that the light rays from a point in the object are parallel. The afocal may be inserted and removed to vary the power and field, or it may be rotated end for end about a central axial pivot point so that the power and field are changed by a factor of MP or 1/MP.

2.6 Bravais System

The Bravais system, shown in Fig. 2.9, can be regarded as the finite conjugate version of an afocal system. The object and image of the Bravais are in the same location, but the size of the image is changed by the linear magnification of the Bravais. In Fig. 2.9 the (virtual) object for the Bravais system is the image formed by the optical system to the left of the figure. The component powers for a Bravais can be found using Eqs. (1.33) and (1.34), setting T equal to zero, so that $T = d+s'-s = 0$. Then the component powers are

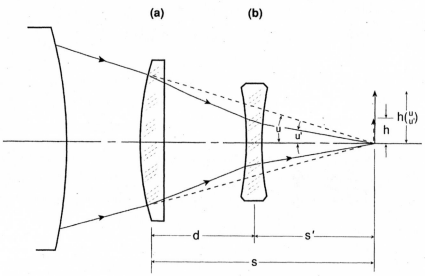

Figure 2.9 A Bravais system does not change the location of an image but does change its size. The Bravais is usually placed between a lens and its image as shown here but could also be placed between the lens and the object.

$$\phi_a = \frac{(m-1)(1-K)}{md} \qquad (2.13)$$

$$\phi_b = \frac{1-m}{d(1-K)} \qquad (2.14)$$

$$K = \frac{d}{s} \qquad (2.15)$$

Note that, for the system shown in Fig. 2.9, the magnification m (equal to u/u') is positive and greater than 1. A positive magnification of less than 1, with the component powers in reverse order from that shown in the figure, is quite possible, but this is a somewhat more difficult system.

Sample calculations

Lay out a 2× Bravais system, 2 in long (i.e., 2 in from lens a to the image). Find the component powers.

With reference to Fig. 2.9, s must be 2 in, and, since $d+s' = s$, we get $d = 2 - s'$ or $s' = 2 - d$. If we assume that $d = 1$ in and thus $s' = 1$ in, K from Eq. (2.15) = 1/2 = 0.5, and Eqs. (2.13) and (2.14) yield

$$\phi_a = \frac{(2-1)(1-0.5)}{2} \cdot 1 = +0.25 \qquad \text{and} \qquad f_a = +4.0$$

$$\phi_b = \frac{1-2}{1(1-0.5)} = -2.0 \qquad \text{and} \qquad f_b = -0.5$$

Since we are free to choose any reasonable value for d, we can try $d = 1.25$ and $s' = 0.75$. Then $K = 1.25/2 = 0.625$ and

$$\phi_a = \frac{(2-1)(1-0.625)}{2} \cdot 1.25 = +0.15$$

$$\phi_b = \frac{1-2}{1.25(1-0.625)} = -2.1333$$

Or if $d = 0.75$ and $s' = 1.25$, then $K = 0.75/2 = 0.375$ and

$$\phi_a = \frac{(2-1)(1-0.375)}{2} \cdot 0.75 = +0.41666$$

$$\phi_b = \frac{1-2}{0.75} \cdot (1-0.375) = -2.1333$$

Thus d can be used as a free variable to control the powers of the components. Now, lay out a Bravais with a power of $0.5\times$ and a 2-in length. Again assuming $d = 1$ in, $K = 0.5$ and

$$\phi_a = \frac{(0.5-1)(1-0.5)}{0.5 \cdot 1} = -0.5$$

$$\phi_b = \frac{1-0.5}{1(1-0.5)} = +1.0$$

2.7 Field Lenses; Relay Lenses; Periscopes

In Fig. 2.10A we show a simple telescope. In order to cover the field of view indicated by the dashed rays, the eyelens would require a very large diameter. In Fig. 2.10B a *field lens* located at the internal image plane converges the light rays at the edge of the field toward the axis so

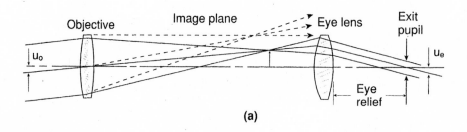

(a)

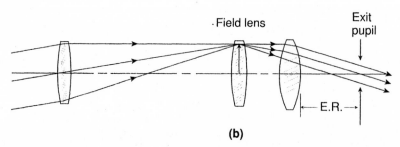

(b)

Figure 2.10 A field lens, located near an image, can redirect the rays at the edge of the field so that they will pass through a small eyelens. The dashed rays in the upper sketch miss the eyelens; in the lower sketch the field lens converges these rays toward the axis so that they go through the eyelens. As indicated, this shortens the eye relief.

that they will pass through a smaller eyelens. Note that this shortens the eye relief; a short eye relief and a small-diameter eyelens go hand in hand. One can lengthen the eye relief with a negative (diverging) field lens; this requires a large eyelens diameter. If A' is the apparent field angle and R is the eye relief, the required diameter of the eyelens (for zero vignetting) is equal to the exit pupil diameter plus $A' \cdot R$. Note that by placing the field lens exactly on the internal image location, the size of the image and the power of the telescope are unchanged. [Using Eq. (1.28), if we set $d = f_{e'}$, we find that $f_{ab} = f_{a'}$.] In actual practice, field lenses are usually located away from the image plane, because otherwise any flaws, scratches, dirt, etc., on the lens would be seen, nicely in focus and magnified by the eyelens. Shifting the field lens away from the focus will cause only a small change in image size and system power.

Sample calculations

In Sec. 2.4 we calculated the component powers and eye relief for a 10-in-long, 4× Kepler telescope as $F_o = +8$ in, $F_e = +2$ in, and a 2.5-in eye relief. If the objective diameter is 1 in and the eyelens diameter is 0.8 in, what power field lens is needed to allow a real field of 1 in diameter (about 7.2°)? What is the eye relief with this field lens?

Assuming that the field lens is located at the internal image, it must bend (converge) the ray from the bottom of the objective ($y = -0.5$ in) to the top of the field lens ($y = +0.5$ in) so that the ray passes through the top of the eyelens ($y = +0.4$ in). This ray climbs 1 in in traveling 8 in from the objective to the focus; its slope is $u = +1.0/8.0 = +0.125$. After refraction by the field lens, it must drop from the top of the field lens ($y = +0.5$ in) to the top of the eyelens ($y = 0.4$ in) in a distance of 2.0 in ($= F_e$); its slope is $u' = -0.1/2.0 = -0.05$.

We can find the necessary field lens power from Eq. (1.19):

$$u' = u - y\phi$$

$$-0.05 = +0.125 - 0.5\phi$$

$$\phi = \frac{-0.05 - 0.125}{-0.5} = +0.35$$

$$f = \frac{1}{0.35} = 2.857 \text{ in}$$

To find the eye relief, we trace the principal ray from the center of the objective with a slope of $u_p = 0.5/8 = +0.0625$, and determine its axial intercept after it passes through the field and eye lenses. The ray-trace is tabulated below.

ϕ	+0.125		+0.35		+0.5	
d		8.0		2.0		
y	0.0		0.5		0.275	
u		0.0625		−0.1125		−0.25

and the eye relief $l_k' = -y_k/u_k' = -0.275/(-0.25) = +1.10$ in.

Sample calculations

If the field lens in the preceding is placed 0.25 in to the right of the internal image (so that its surfaces are out of focus), what clear aperture and power will be required for the field lens? What eye relief will result?

Using the same ray from the bottom of the objective to the top of the 1-in-diameter field, the additional travel distance of 0.25 in at a slope of +0.125 will increase the ray height at the field lens by $0.25 \cdot 0.125 = 0.03125$ to $y = 0.53125$. The slope to make this ray strike the eyelens at $y = 0.4$ is then $(0.4-0.53125)/(2.0-0.25) = -0.075$. Again, using Eq. (1.19),

$$u' = u - y\phi = -0.075 = 0.125 - 0.53125\phi$$

$$\phi = \frac{-0.075-0.125}{-0.53125} = +0.376471$$

$$f = 2.656247 \text{ in}$$

The raytrace to determine the eye relief is tabulated below.

ϕ	+0.125		+0.376471		+0.50	
d		8.25		1.75		
y	0.0		0.515625		0.285294	
u		+0.0625		−0.131618		−0.274265

and the eye relief is $l_k' = -0.285294/(-0.274265) = 1.040213$. But note that if we move the field lens away from the focus point, it will throw the telescope out of focus. We can choose to maintain either the telescope power or its length. If we wish to maintain the 10-in length, we will have to change the eyepiece focal length. The focal length and back focus of the combined objective and field lens are found from Eqs. (1.28) and (1.29):

$$f_{ab} = \frac{f_a f_b}{f_a + f_b - d} = \frac{8 \cdot 2.656}{8 + 2.656 - 8.25} = 8.831170$$

$$B = \frac{f_{ab}(f_a - d)}{f_a} = \frac{8.831(8 - 8.25)}{8} = -0.275974$$

Thus, to make the focal points of objective and eyelens coincide, the eyelens focal length must equal 1.75 + 0.275974 = 2.025974 in. Again raytracing to determine the power and eye relief, we get

ϕ		+0.125		+0.376471		+0.493590	
d			8.25		1.75		
y		0.0		0.515625		0.285294	
u	+0.0625		+0.0625		−0.131618		−0.272436

The eye relief is $l' = -y/u' = -0.285294/(-0.272436) = 1.047196$, very little changed. But the telescope power, per Eq. (2.1), has increased to MP $= (A'/A) = -0.272436/0.0625 = -4.359$. If we wish to maintain the power at MP $= -4\times$, the eyepiece focal length must be $8.831/4 = 2.207792$, and the telescope length becomes

$$L = 8 + 0.25 - 0.275974 + 2.207792 = 10.181818$$

If we needed *exactly* $4\times$ power and 10-in length, we could use one of the techniques described in Chap. 5 to find a simultaneous solution.

A *relay lens* is simply a lens which relays an image from one point to another. For example, the erector lens in Fig. 2.7 is also a relay lens.

A *periscope* is typically a train of alternating relay and field lenses which are arranged to transmit an image through a long narrow space. Usually the desired field of view is larger than can be passed through the available space. Figure 2.11 shows a schematic periscope. The layout which produces the minimum number of components and the (desirable) minimum amount of power for a given length and diameter can be found as follows: The initial objective focal length is chosen so that the image of the desired field of view 2α just fills the available diameter D_2; thus $2 \cdot \alpha \cdot F_o = D_2$, or $F_o = D_2/2\alpha$. The distance S_1 to the first relay lens is chosen so that the cone of light from the objective just fills the available diameter for the relay; thus $(D_1/F_o) = (D_3/S_1)$, or $S_1 = D_3 \cdot F_o/D_1$. The power of the field lens B is chosen to focus the image of lens A in lens C. Relay lens C images field lens B on field lens D so that the image fills the aperture of D; thus $(D_2/S_1) =$

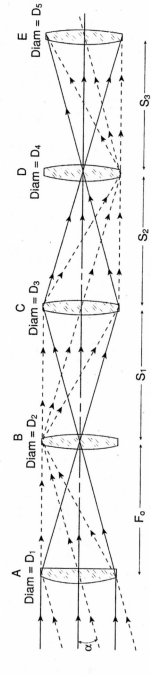

Figure 2.11 Periscope schematic, showing the objective lens A forming its image at the field lens B which images the objective in the first relay lens C. The relay lens C forms its image in field lens D which images the pupil in relay lens E, etc. This technique is used to pass an image through a long narrow space. The periscope can be terminated with an eyepiece, a camera, or other optical instrument. Submarine periscopes, borescopes, and medical endoscopes all use this principle of alternating relay and field lenses.

(D_4/S_2). The distance to relay lens E is again chosen to fill the aperture of E; thus $(D_3/S_2) = (D_5/S_3)$ and field lens D images lens C on lens E. This process can be continued as necessary, to the final eyepiece or camera focal plane. It should be obvious that, if all the diameters are equal ($D_1 = D_2 = D_3 = D_4 =$ etc.), the relay lenses (C, E, etc.) will work at unit magnification ($m = -1$), as will the field lenses (D, F, etc.), and the focal lengths of lenses C, D, E, F, etc., will be identical.

Sample calculations

A periscope is to have a maximum clear diameter of 4 in for the optics and it is to cover a (paraxial) field of view of ± 0.1. Assume the objective diameter is 2 in and all other optics are 4 in diameter. Lay out a system with minimum component power.

The maximum image distance for the objective is equal to the maximum optics diameter divided by the full field angle, or 4 in/(2•0.1) = 20 in. (Note that this applies whether the object is at infinity or some finite distance, and it produces the lowest possible power objective.) The objective image side NA is then equal to its semidiameter divided by the image distance, or 1/20 = 0.05, corresponding to a "working" speed of $f/10$. In order to fill the 4 in diameter of the first relay lens, its distance from the objective focus is 4 in$\times f/\# = 40$ in, and the focal length of the first field lens is determined from Eq. (1.4):

$$\frac{1}{s'} = \frac{1}{f} + \frac{1}{s} = \frac{1}{40} = \frac{1}{f} + \frac{1}{-20}$$

$$\frac{1}{f} = 0.075 \qquad f = 13.333...$$

The relay lens will work at unit magnification, and its conjugate distances are $s = -40$ and $s' = +40$. Its focal length per Eq. (1.4) is thus 20 in. The second field lens also works at unit magnification and 40-in conjugate distances. Its focal length is 20 in, as are the focal lengths of all subsequent field and relay lenses. There will be images located at distances from the objective of 20, 100, 180, 260 in, etc. Since the relay stages are all of unit magnification, the complete system will have a focal length equal to that of the objective lens, or ($\pm$)20 in. The optics to be used at and after the final image will depend on the desired final image size and whether the device is to be a telescope, a camera, or another type of instrument.

2.8   Magnifiers and Microscopes

The magnification of a microscope or a magnifier is, like that of a tele-scope, defined as the ratio of the angle subtended by the image to the angle subtended by the object. The difficulty here is that we are concerned with an object at a finite distance, and the angle which the object subtends will vary with that distance. The answer to this dilemma is that the object is considered to be viewed at a conventional distance of 10 in, chosen as the "nearest distance of distinct vision." This convention is obviously a compromise, since the eyes of a young person can focus to a distance of a few inches, whereas an older individual may be unable to focus closer than several feet.

If the object to be examined is placed at the focal point of the microscope/magnifier, the image is seen at infinity, and the magnification is simply

$$MP = \frac{10 \text{ in}}{F} = \frac{250 \text{ mm}}{F} \qquad (2.16)$$

This expression is valid for either a simple magnifying glass or a compound microscope, where F is the effective focal length of the microscope/magnifier. If, as shown in Fig. 2.12, the object is between the lens and the first focal point, then the magnification depends not only on the focal length F, but also on the image distance S', as well as the distance from the magnifier to the eye R, as follows:

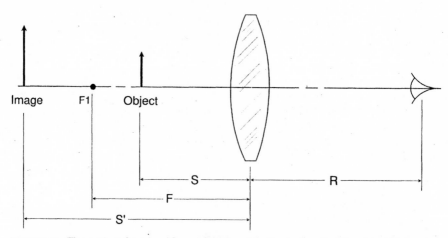

Figure 2.12   The optics of a magnifier or simple microscope, showing the object distance S, the image distance S', and the eye distance R. The object is located at, or within, the first focal point F_1, and the image is virtual and located far enough from the eye that it can be comfortably viewed; this image distance is usually quite large.

$$MP = \frac{10 \text{ in}(F - S')}{F(R - S')} \tag{2.17}$$

Equation (2.16) is used to determine the stated power of magnifiers, eyepieces, and microscopes, whereas Eq. (2.17) is useful for determining the magnification of devices such as slide viewers. It should be apparent that if the definition of magnification were changed to use the angle subtended by the object from a distance D instead of 10 in, we would simply substitute D for 10 in in the above equations.

Sample calculations

Lay out a 4× tabletop slide viewer which is to be used at a distance of 20 in from the eye, and which is to provide an image that is -0.67 diopters from the eye. A distance of -0.67 diopters is a distance of $(1/-0.67) = -1.5$ m ≈ -60 in. With reference to Fig. 2.12, $R = 20$ in and $S' = -40$ in. Using Eq. (2.17),

$$MP = \frac{10(F - S')}{F(R - S')}$$

$$4\times = \frac{10[F - (-40)]}{F[20 - (-40)]} = \frac{10(F + 40)}{60F}$$

$$240F = 10F + 400$$

$$F = \frac{400}{230} = 1.7391 \text{ in}$$

To locate the slide position, we use Eq. (1.4):

$$\frac{1}{S'} = \frac{1}{S} + \frac{1}{F}$$

$$\frac{1}{-40} = \frac{1}{S} + \frac{1}{1.7391}$$

$$S = -1.66666 \text{ in}$$

For a slide diagonal of 1.6 in, how large must the lens diameter be? The magnified image is $4\times1.6 = 6.4$ in, and at an image distance of 40 in it subtends an angle of $6.4/40 = 0.16$. At the eye distance of 20 in this angle requires a lens diagonal of $20\times0.16 = 3.2$ in. This is a large diameter for a single lens element with a focal length of only 1.74 in (as a quick sketch will show; see Sec. 5.5 for sketch techniques). At this point our alternatives are: (1) change the initial specifications so that the lens focal length is longer and its required diagonal is smaller; (2) use more than one element for the lens; (3) make the lens smaller than 3.2 in and force users to shift their heads to see the full slide; or some combination of these.

In the *compound microscope,* shown in Fig. 2.13, the objective lens forms a magnified image of the object, which is viewed through the eyepiece. The magnification is the product of the objective magnification (S'/S) and the eyepiece magnification $(10 \text{ in}/F_e)$, or

$$\text{MP} = \frac{S'}{S} \cdot \frac{10}{F_e} \tag{2.18}$$

The microscope magnification can also be determined by calculating the effective focal length of the combination of the objective and eyepiece, using Eq. (1.27) or (1.28). This gives the focal length for the microscope F_m:

$$F_m = \frac{F_e F_o}{F_e + F_o - d}$$

$$= \frac{F_e F_o}{F_e + F_o - F_e - S'}$$

$$= \frac{F_e F_o}{F_e - S'} \tag{2.19}$$

and a magnification of

$$\text{MP} = \frac{10 \text{ in}}{F_m} = \frac{10 \text{ in}(F_o - S')}{F_e F_o} \tag{2.20}$$

which yields exactly the same result as does Eq. (2.18).

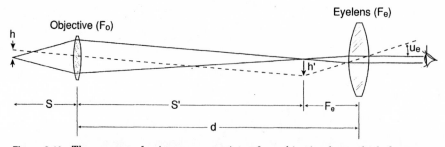

Figure 2.13 The compound microscope consists of an objective lens which forms an enlarged image of the object, and an eyepiece which further magnifies the image and allows the eye to view it comfortably.

Sample calculations

We wish to view an object 40 in away at a magnification of 5×. The optical instrument can be 10 in long. A 5× magnifier/microscope has a focal length of 10/MP = 2 in, but since MP may be plus or minus, so may the focal length. If the object is placed at the focal point of the microscope (so that the image presented to the eye is at infinity for comfortable viewing), the back focus must be $B = 40$ in. The component powers will be minimized if the space between components is the maximum allowed, or $d = 10$ in. Using Eqs. (1.31) and (1.32) and $f_{ab} = +2$ in, we get

Eq. (1.31)

$$f_a = \frac{df_{ab}}{f_{ab} - B}$$

$$= \frac{10 \cdot 2}{2 - 40} = \frac{20}{-38} = -0.5263$$

Eq. (1.32)

$$f_b = \frac{-dB}{f_a - B - d}$$

$$= \frac{-10 \cdot 40}{2 - 40 - 10} = \frac{-400}{-48} = +8.333$$

This is the galilean version, with a negative eyelens and a positive objective; the image is erect; there is no internal focus. If we use MP $= -5×, f_{ab} = -2.0$ in, and we get

$$f_a = \frac{10(-2)}{-2 - 40} = \frac{-20}{-42} = +0.4762$$

$$f_b = \frac{-10 \cdot 40}{-2 - 40 - 10} = \frac{-400}{-52} = +7.692$$

This arrangement corresponds to the conventional compound microscope with an inverted image and an internal focus. The internal image allows the use of a reticle or crosshair.

The optics of a HUD (or *head-up display*) and those of an HMD (or *head/helmet-mounted display*) are basically magnifiers. Typically a HUD projects collimated information from a CRT (cathode ray tube), an LCD (liquid crystal display), or an image intensifier into the eye.

The object is placed at or near the focal point of the optics so that the image produced is at infinity or a large distance. Depending on the application, the infinitely distant image may be reflected from a 45° tilted semireflecting "combiner" mirror, through which the user can also see directly. A military aircraft HUD may provide weapon or aircraft status information and also provide an aiming point for the weapon system. Because the user's eye is usually located a significant distance from the lens system, the field of view of a HUD is limited by the aperture of the optics. The system must also be large enough to accommodate a certain amount of motion of the eye without vignetting (obscuring) the image. In an HMD the eye is usually much closer to the optics and the function is more like that of an ordinary magnifier. Some HMDs have optical systems which incorporate relay optics and are thus analogous to the compound microscope. This is usually done in order to make the center of gravity of the total HMD assembly roughly coincident with that of the head. A concave mirror, either fully or semireflecting is sometimes used as the collimating/magnifier in conjunction with a 45° semireflecting combiner mirror. In some applications, such as an HMD for surgery or virtual reality, there is no see-through capability and the view of the outside world, if required, is seen by looking beneath the HMD optics.

2.9 Telephoto and Retrofocus Arrangements

A schematic *telephoto* arrangement is shown in Fig. 2.14. It consists of a positive component separated from a negative component. The focal length, back focus, etc., relationships are defined by the equations of Sec. 1.10. As can be seen from the figure, the advantage of the telephoto arrangement is that its focal length is longer than its overall length. One can get a long focal length lens in a short, compact package. The *telephoto ratio* is the length L divided by the effective focal length F, and a lens is classed as a true telephoto if this ratio is less than 1. In reflecting systems, the Cassegrain configuration, as shown in Fig. 1.16, is the mirror equivalent of a telephoto.

A schematic *retrofocus* or *reversed telephoto* system is shown in Fig. 2.15. It consists of a negative component followed by a positive component. The advantage of this arrangement is that the back focus distance is long compared to the focal length. This is useful if a long working distance is needed in order to accommodate a mirror, prism, or beam splitter in the space between the optics and the image. Occasionally the retrofocus is executed by placing a negative component at the first focal point of a positive lens; this increases the back

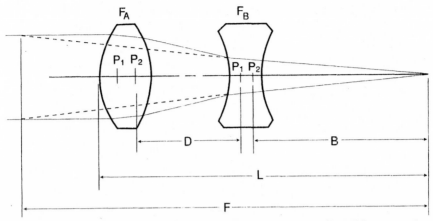

Figure 2.14 A telephoto lens consists of a positive component followed by a negative component. The combination has a focal length F which is longer than its overall length L and is used where a long focal length in a compact package is desired. The telephoto ratio is L/F; a ratio of less than 1 is considered the defining characteristic of a telephoto.

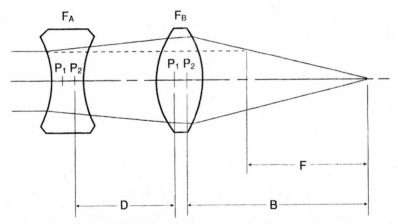

Figure 2.15 A reversed telephoto or retrofocus lens has a negative component followed by a positive component. It provides a long working distance B for a short focal length and is used where it is necessary to put a beam splitter or the like between the lens and its image.

focus or working distance without changing the focal length of the original positive lens. The mirror equivalent of the retrofocus is shown in Fig. 2.23B. A "fisheye" lens covers an extremely wide field by utilizing one or more large, strong meniscus negative elements.

2.10 Collimators

A collimator is simply a device which produces an image at infinity. This is done by placing the object at the focal point of the optical system. The collimator is a common laboratory device, useful when a system designed to be used with a very distant object is tested. A collimated beam of light is one in which a small source is imaged at infinity. A common misconception is that all the light rays in a collimated beam are parallel to each other and that the beam does not expand. This is of course incorrect; the beam does spread, and the angle by which it spreads is equal to the size of the source divided by the effective focal length of the collimating lens. The rays from a single *geometric point* in the object are indeed parallel, but bear in mind that a true geometric point has dimensions of zero by zero, and an area of zero emits no energy. Even a perfect laser source beam will spread because of diffraction. A laboratory collimator is typically a well-corrected lens, such as an achromatic doublet, an apochromatic triplet, or a parabolic mirror, with an illuminated target placed at its focus. A collimated beam is often used in applications where a minimum beam spread is desired.

2.11 Anamorphic Systems

An anamorphic system is one which has a different magnification or focal length in each of the prime meridians. This is usually accomplished with lens elements whose surfaces are cylindrical (or toric). Figure 2.16 shows a system consisting of two cylindrical elements with

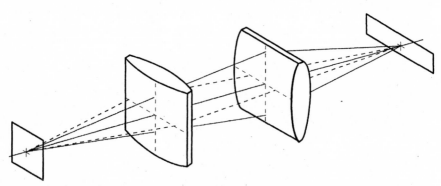

Figure 2.16 An anamorphic system has a different focal length or magnification in each of the prime meridians. Here the two cylindrically surfaced components produce a system with a magnification of about 0.5× vertically and about 2.0× horizontally, so that the image of the square object at the left is a rectangle at the right (with a 4:1 aspect ratio).

their power axes at right angles to each other. The first element focuses the rays of a horizontal fan (shown dashed) to produce a magnified image of the square object; for rays in a vertical (meridional) fan this first element behaves as a plane-parallel plate and does not deviate these rays. The second element will focus the meridional rays, producing an image smaller than the object; the horizontal section of this element is plane-parallel. The image of the square object is a rectangle. Because the power of a cylinder varies as the square of the cosine of the angle that a ray fan makes to the power meridian of the lens, it turns out that if the anamorphic system is in focus in both prime meridians, it is in focus in all meridians. This means that a sharp (if anamorphically distorted) image can be produced by such a system.

Sample calculations

The output of a laser beam is 20 mm high and 2 mm wide. Image it as a 2-mm square at a distance of 2 ft. We simply handle the imagery in each meridian separately. The 2-mm width must be imaged at unit magnification. The cylinder lens (with a vertical cylinder axis) must be located midway between object and image so that $s' = -s = 12$ in. Equation (1.4) can be solved for the necessary focal length of 6 in. In the vertical meridian we need a magnification of $-2/20 = -0.1$. The object and image distances must be in a ratio of 10 to 1, and they must add up to 24 in (2 ft). Thus $s = -10s'$ and $s'-s = 24$. Substituting, we get $s'+10s' = 24$, so that $s' = 24/11 = 2.1818$ in and $s = -10s' = 21.8181$ in. Again using Eq. (1.4), we get a focal length for the second cylinder lens (with horizontal cylinder axis) of 1.9835 in.

Probably the most widely used anamorphic device is that used to produce wide-screen motion pictures. As shown in Fig. 2.17, a cylindrical afocal attachment in the form of a reversed (0.5×) galilean telescope is placed in front of the camera lens with its cylinder axes vertical. The combined camera lens and anamorphic attachment has a focal length equal to that of the camera lens for rays in the vertical meridian, but equal to half that focal length for rays in the horizontal meridian. The result is that the picture on the film is compressed horizontally by a factor of 2, enabling a wide horizontal field to be imaged on standard 35-mm film in a standard camera. When the film is projected in the theater, a similar attachment is placed before the projection lens, the projected image is restored to its original proportions, and a wide-screen picture results.

A difficulty with such a camera lens is that it has a different focal length in each meridian; the required movement of a lens to focus on

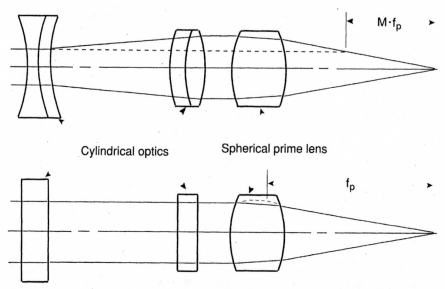

Figure 2.17 An anamorphic system can be produced by combining ordinary spherical-surfaced optics with an anamorphic attachment. Here an afocal anamorphic attachment produces a lens whose focal length in the meridian of the upper sketch is one half of that in the lower. This arrangement is used in wide-screen movies to get a wide-aspect-ratio picture using a standard camera and film.

nearby subjects is four times greater in one meridian than in the other. (Newton's equation, Eq. (1.1), $x' = -f^2/x$, gives the required lens motion as x'.) The obvious way to handle this is to focus the prime lens independently in the vertical meridian (where the attachment has zero power) and then focus the anamorphic attachment in the horizontal meridian by changing the space between its positive and negative components. The difficulty with this is that the spacing adjustment also changes the magnification of the anamorphic attachment so that, for example, faces appear fatter in closeups than in actuality. This is not a popular solution among the members of the acting profession. A more acceptable solution is to add a pair of weak spherical elements, one positive and one negative, in front of the anamorph as shown in Fig. 2.18. The element powers are chosen such that when they are closely spaced the combination has zero power, but when the spacing is increased they have positive power. This arrangement can focus the camera by "collimating" the object (i.e., presenting to the anamorphic system an image of the object which is located at infinity) without changing the anamorphic ratio.

A Bravais system, as discussed in Sec. 2.5, can be executed with cylindrical surfaces and used behind a camera lens as an anamorphic

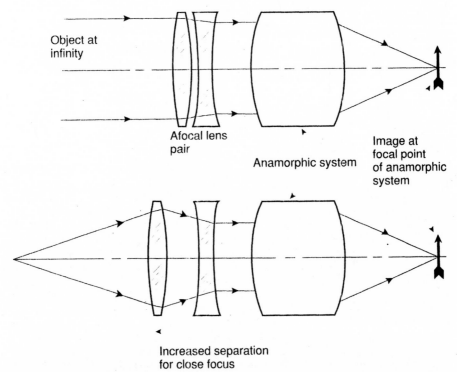

Object at
infinity

Afocal lens
pair

Anamorphic system

Image at
focal point
of anamorphic
system

Increased separation
for close focus

Figure 2.18 This shows a focusing device which is useful for problem systems such as anamorphic lenses and zoom lenses, where the shift to focus the lens is different for each meridian, or changes with the zoom setting. The pair of weak elements has zero power when closely spaced, but positive power when separated. They "collimate" the object, so that the anamorph or zoom lens is always presented with an image of the object at infinity, and no focus shift of the main lens is necessary.

attachment. The advantage to this arrangement is that the Bravais attachment may be much smaller than the galilean, and this smaller size is significant when the prime lens is a long-focal-length, large-diameter lens. Note that, since the Bravais increases the image size and the reversed galilean reduces the size, the cylinder axes of the Bravais must be horizontal rather than vertical to produce the same anamorphic effect. Also, the system may be focused by moving just the prime lens; this avoids the problem mentioned in the preceding paragraph.

As shown in Fig. 2.19, an ordinary refracting prism is an afocal anamorphic system. The parallel rays of the incoming and emerging ray bundles indicate that the device is afocal with object and image at infinity. The different widths of the ray bundles indicate that there is an angular magnification in the meridian shown in the figure. In the

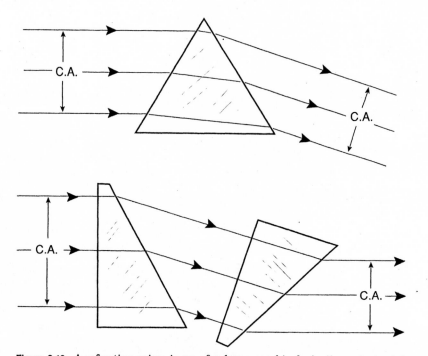

Figure 2.19 A refracting prism is an afocal anamorphic device (except at minimum deviation) as indicated by the parallel input and output rays. The ratio of the ray separations indicates the angular magnification in the meridian of the page; this ratio is a function of the angle at which the rays enter the prism. The lower sketch shows two prisms arranged so that their deviations cancel each other but their anamorphic effects multiply.

other meridian, the prism appears as a plane-parallel plate and has no angular magnification. Both the angular deviation of the beam and the anamorphic angular magnification vary as the angle of incidence on the prism is changed. As shown in the lower sketch, a pair of prisms can be arranged as shown so that their deviations and dispersions cancel and their anamorphic effects multiply. If the prisms are achromatized, this device can be used in the same way that the galilean projection anamorph described above is used, although its aberrations limit it to smaller angular fields. A suitably synchronized rotation of both prisms will allow the anamorphic ratio to be varied.

An LED (light-emitting diode) has an emission characteristic which is awkward to collimate, in that its beam spreads out more widely in one meridian than in the other, typically by a factor of about 3. It also has astigmatism, in that the beam appears to originate and spread out from two longitudinally separated points. A collimator for such an LED is shown in Fig. 2.20. The elliptical beam spread is made circu-

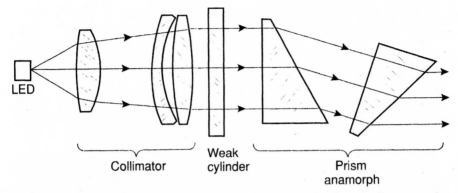

Figure 2.20 A laser diode collimator consisting of a spherical collimating lens, a weak cylindrical lens to correct the astigmatism of the LED, and a prism anamorph to convert the elliptical output beam of the LED to a circular beam.

lar by a simple prismatic anamorph (which need not be achromatized for this application), and the astigmatism is corrected by a weak cylindrical surface, producing a well-collimated circular beam.

2.12 Zoom and Varifocal Systems

Both zoom and varifocal systems are characterized by the ability to change focal length by longitudinally sliding one or more components with respect to the balance of the system. The "zoom" system implies a fixed focal plane, as in a video or movie camera lens. The "varifocal" has a variable focal length; the location of the focal point may vary. For example, some still camera or projection lenses require refocusing when the focal length is changed.

The simplest way to change focal length is to change the spacing between two components. The equations of Sec. 1.10, specifically Eqs. (1.27) to (1.30), can be used to determine the focal length change produced by respacing a two-component system. Most two-component zooms consist of one positive and one negative component. Both are moved to simultaneously change the focal length and maintain the focus. If the negative component leads, the system resembles the reverse telephoto (or retrofocus) described in Sec. 2.8, Fig. 2.15. This arrangement yields a short focal length and a long working distance, plus the capability for a wide field of view. The reverse, with a positive component followed by a negative component, resembles the telephoto (see Fig. 2.14) and provides a long focal length in a short package. It is obviously also possible to make a zoom with two positive components, but if any significant field of view is to be covered, off-axis image quality considerations strongly favor systems with both positive and negative components.

Figure 2.21 shows the motion of the components for both types. Here we have arbitrarily assumed that the component focal lengths are +1.0 and −1.25. Note that the range of the retrofocus type is much greater because the telephoto type runs out of back focus.

A three-component zoom is also a common arrangement and often consists of a negative component sliding between two positive components, one of which may be moved to maintain focus. The added flexibility provided by three components allows a wide variety of arrangements, including afocal or telescope systems. Several three-component zoom schematics are shown in Fig. 2.22. Some modern zoom camera lens designs have three or more moving components; this is done to allow the image quality to be consistently good over a large zoom range.

Note that the zoom lens has a focusing problem similar to that of an anamorphic system, as described in the preceding section. The lens motion required to focus on a close object will vary as the square of the focal length. One solution is the same as that sketched in Fig. 2.18. In practice, most zoom lenses have multielement components and can be focused by changing the spacing between the elements within a component (usually the front component).

Another complication of zoom lenses is that, unless the aperture stop (or iris diaphram) is located after the sliding components (i.e., between the last moving element and the image), the relative aperture (f-number or numerical aperture), and hence the image illumination, will vary when the lens is zoomed. In the newer autofocus and autoexposure cameras, the automatic adjustment of focus and exposure makes these considerations far less important.

2.13 Mirror Systems

It should be apparent that the equations in Chap. 1 and the systems in the preceding sections apply to and can be executed using mirrors. We have previously noted that (1) the focal length of a mirror is one-half of its radius, (2) a concave mirror has positive power, and (3) the principal points are at the mirror surface. But when raytracing mirrors, there is a bit of complexity. For surface-to-surface raytracing [Sec. 1.5; Eqs. (1.14) and (1.15)] the spacing is negative if a following surface is to the left, and the index is negative when the ray travels from right to left (i.e., as after a reflection). However, for component-by-component raytracing [Sec. 1.7; Eqs. (1.19) and (1.20)], we use the "air-equivalent distance," which is simply the distance divided by the

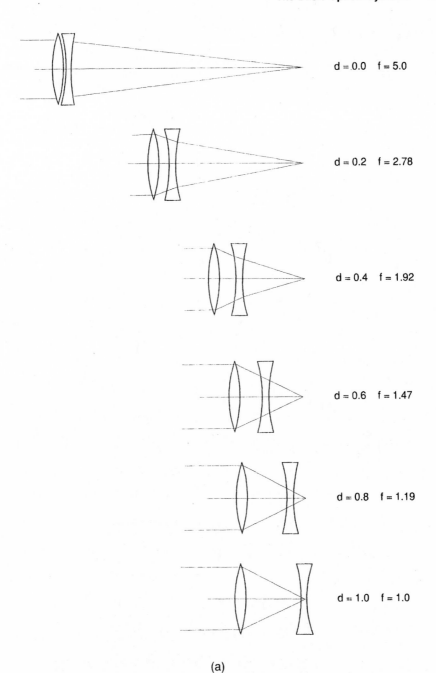

$d = 0.0$ $f = 5.0$

$d = 0.2$ $f = 2.78$

$d = 0.4$ $f = 1.92$

$d = 0.6$ $f = 1.47$

$d = 0.8$ $f = 1.19$

$d = 1.0$ $f = 1.0$

(a)

Figure 2.21 Schematic of two two-component zoom systems, showing the motions necessary to change the focal length and maintain the image focus. The component focal lengths are $+1.0$ and -1.25. (A) With the positive lens in front, the useful zoom range is limited by the lack of back focus.

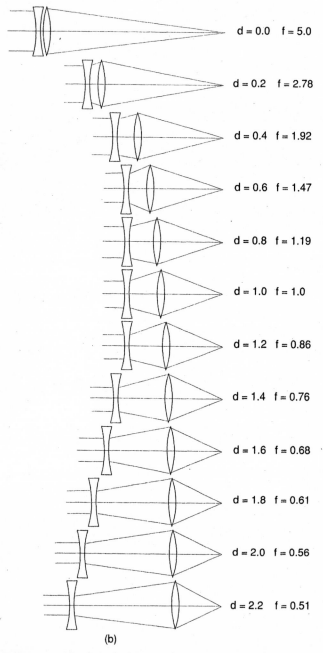

d = 0.0   f = 5.0

d = 0.2   f = 2.78

d = 0.4   f = 1.92

d = 0.6   f = 1.47

d = 0.8   f = 1.19

d = 1.0   f = 1.0

d = 1.2   f = 0.86

d = 1.4   f = 0.76

d = 1.6   f = 0.68

d = 1.8   f = 0.61

d = 2.0   f = 0.56

d = 2.2   f = 0.51

(b)

Figure 2.21 *(Continued)* *(B)* The reversed arrangement allows both a greater zoom range and a good back focus clearance distance.

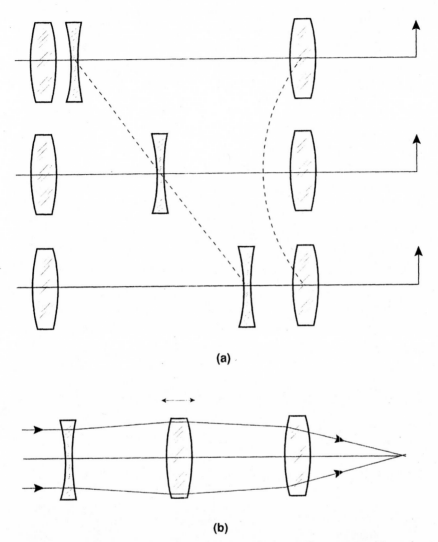

(a)

(b)

Figure 2.22 Three-component zoom systems. (A) A negative component moving between two positive components to change the focal length is a very common arrangement. Either of the positive components may be moved to compensate for the focus shift introduced by moving the inner lens. (B) A moving positive lens can also produce a zoom.

index. Thus, when both are negative, we take the spacing as positive. This is the case for most two-mirror systems. This rule applies not only to component-by-component raytracing but also to the equations for combinations of two components [Sec. 1.10, Eqs. (1.27) through (1.37)] when applied to mirror systems.

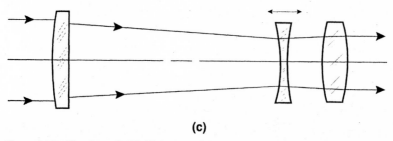

(c)

Figure 2.22 (Continued) (*C*) This is an afocal zoom which can be used as an attachment to make a zoom projection lens from an ordinary fixed-focus lens. It goes out of focus, but in use it can be refocused after the picture size is adjusted.

With that formality out of the way, we can discuss a number of popular two-mirror systems. Remembering to use a positive sign for d (per the paragraph above), and using the same notation as in Eqs. (1.27) to (1.37), we have the following equations for two-mirror systems:

$$f_{ab} = \frac{r_a r_b}{2(r_a + r_b - 2d)} \tag{2.21}$$

$$B = \frac{f_{ab}(r_a - 2d)}{r_a} \tag{2.22}$$

where r_a and r_b are the mirror radii, d is the spacing, B is the back focus, and f_{ab} is the focal length of the combination. Note that we use the usual sign convention for the mirror radii; i.e., r is positive if the center of curvature is to the right of the surface.

To determine the mirror radii, given the combined focal length, the back focus, and the spacing, we can use

$$c_a = \frac{1}{r_a} = \frac{B - f_{ab}}{2df_{ab}} \tag{2.23}$$

$$c_b = \frac{1}{r_b} = \frac{B + d - f_{ab}}{2dB} \tag{2.24}$$

where c is the surface curvature, equal to $1/r$.

Several of the commonly encountered two-mirror arrangements are diagramed in Fig. 2.23. Figure 2.23*A* is called a Cassegrain and is the mirror equivalent of the telephoto arrangement, where a long focal length is produced in a compact package. The Cassegrain is perhaps the most widely used of all the two-mirror systems. Figure 2.23*B* shows what is called a Schwarzschild; it is the mirror analog of the

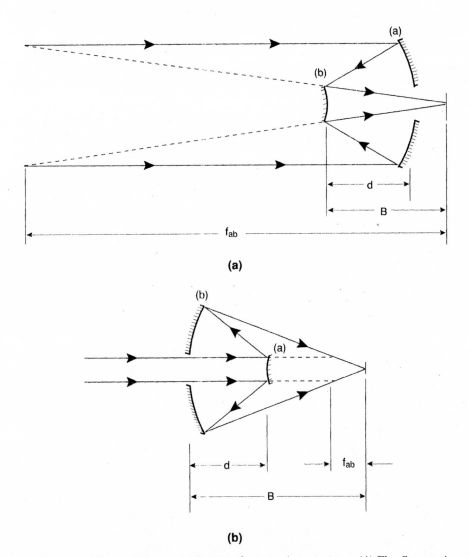

(a)

(b)

Figure 2.23 Three common arrangements for two-mirror systems: (A) The Cassegrain objective with a concave primary mirror and a convex secondary is a compact system with a long focal length, in a sort of telephoto configuration. The most commonly used arrangement. (B) The Schwarzschild system has a convex primary and a concave secondary, producing a long working distance and a short focal length (at the expense of a secondary mirror diameter which is several times the beam diameter). Often found in short-focal-length microscope objectives for ultraviolet and infrared work.

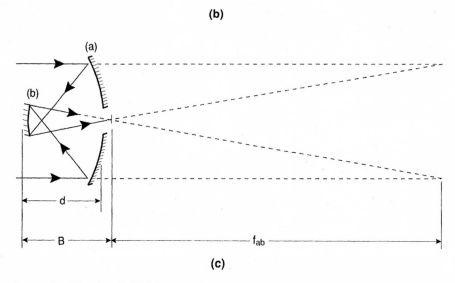

(c)

Figure 2.23 (*Continued*) (*C*) The gregorian arrangement uses a concave secondary mirror to relay the image through a hole in the primary and erect the image. Rarely used.

retrofocus, since it has a long working distance compared to its focal length. It is often used as an infrared or ultraviolet microscope objective. The gregorian arrangement (Fig. 2.23*C*) is the mirror equivalent of a positive component forming an image which is then relayed and magnified by a relay lens. The focal length f_{ab} of the gregorian is negative, since P_2 is to the right of F_2. Mirror systems have the advantage that a mirror has no chromatic aberration and that they do not require high-quality optical materials to transmit the wavelengths of interest.

Note that in all of these systems, the central part of the beam is obscured by one of the mirrors, and the other mirror has a central hole for the beam to pass through. The Cassegrain and the gregorian are usually executed with aspheric surfaces on both mirrors (to correct aberrations). For the Schwarzschild, the aspherics are not necessary; it can be made from two spheres. As indicated in Chap. 4, the central obscuration of the beam has the diffraction effect of significantly reducing the image contrast.

Afocal mirror systems are shown in Fig. 2.24. A common arrangement is to use "confocal" paraboloid mirrors as shown in Figs. 2.24*A* and *B*. These are of course the mirror equivalents of the galilean and keplerian telescopes, respectively. Some mirror systems use an off-axis aperture stop in order to avoid obscuring the center of the beam. Figure 2.24*C* and *D* shows this arrangement for confocal paraboloids, and Fig. 2.24*E* shows what is often mistakenly called an "off-axis

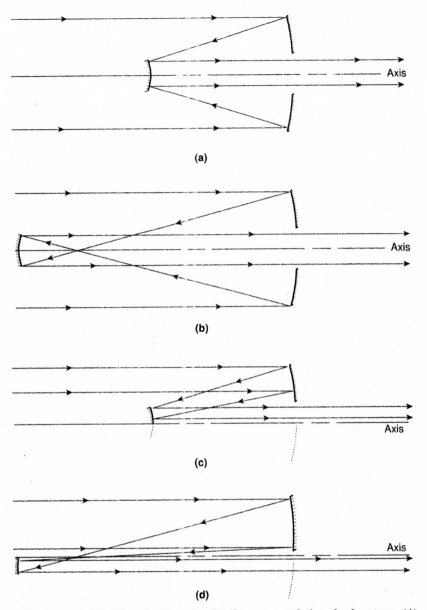

(a)

(b)

(c)

(d)

Figure 2.24 (A), (B), (C), and (D) are "confocal" systems and also afocal systems. (A) and (B) are the mirror analogs of the galilean and Kepler telescopes, respectively. (C) and (D) are the same except that the apertures are decentered in order to avoid an obscuration in the center of the beam. These four are often made with parabolic surfaces to achieve a system without spherical, coma, or astigmatism.

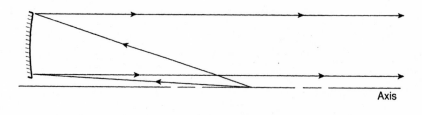

(e)

Figure 2.24 (*Continued*) (*E*) is an "off-axis" paraboloid used as a collimator. Note that the mirror is used on-axis; the aperture is what is off-axis.

parabola" (the *aperture* is off-axis, not the paraboloid) used as a collimator.

2.14 Collected Equations

Afocal systems

Figure 2.4
Eq. (2.1)

$$MP = \frac{A'}{A} = \frac{P}{P'}$$

Eq. (2.2)

$$MP = \frac{\tan(A'/2)}{\tan(A/2)} = \frac{P}{P'}$$

Eq. (2.3)

$$m = \frac{P'}{P} = \frac{1}{MP}$$

Eq. (2.4)

$$MP = -\left(\frac{F_o}{F_e}\right)$$

Keplerian telescopes

Figure 2.5
Eq. (2.5)

$$L = F_o + F_e$$

Eq. (2.6)

$$R = \frac{(MP - 1)F_e}{MP}$$

Eq. (2.7)

$$F_o = \frac{MP \times L}{MP - 1}$$

Eq. (2.8)

$$F_e = \frac{L}{1 - MP}$$

Galilean telescope

Figure 2.6

Eq. (2.9)

$$A' = \frac{F_o}{L(f/\#)}$$

Eq. (2.10)

$$A = \frac{F_o}{L(f/\#)MP} = \frac{-F_e}{L(f/\#)}$$

Terrestrial telescope

Figure 2.7

Eq. (2.11)

$$MP = -\left(\frac{F_o}{F_e}\right)\left(\frac{S'}{S}\right)$$

Afocal attachments

Figure 2.8

Eq. (2.12)

$$F_c = MP{\cdot}F_p$$

Bravais system

Figure 2.9

Eq. (2.13)

$$\phi_a = \frac{(m - 1)(1 - K)}{md}$$

Eq. (2.14)

$$\phi_b = \frac{1 - m}{d(1 - K)}$$

Eq. (2.15)

$$K = \frac{d}{s}$$

Magnifiers
Figure 2.12
Eq. (2.16)

$$MP = \frac{10 \text{ in}}{F} = \frac{250 \text{ mm}}{F}$$

Eq. (2.17)

$$MP = \frac{10 \text{ in}(F - S')}{F(R - S')}$$

Compound microscopes
Figure 2.13
Eq. (2.18)

$$MP = \frac{S'}{S} \cdot \frac{10 \text{ in}}{F_e}$$

Eq. (2.19)

$$F_m = \frac{F_e F_o}{F_e - S'}$$

Eq. (2.20)

$$MP = \frac{10 \text{ in}(F_o - S')}{F_e F_o}$$

Mirror systems
Figure 2.23
Eq. (2.21)

$$f_{ab} = \frac{r_a r_b}{2(r_a + r_b - 2d)}$$

Eq. (2.22)

$$B = \frac{f_{ab}(r_a - 2d)}{r_a}$$

Eq. (2.23)

$$c_a = \frac{1}{r_a} = \frac{B - f_{ab}}{2df_{ab}}$$

Eq. (2.24)

$$c_b = \frac{1}{r_b} = \frac{B + d - f_{ab}}{2dB}$$

3

Condensers, Illuminators, Photometry, Etc.

3.1 Interchangeability of Sources and Detectors

This chapter is written as a discussion of illuminating systems, based on the concept that we want to efficiently produce some level of uniform illumination from a given source. In almost every case, however, one can substitute a detector for the source and reverse the direction of the light, and the result is an analogous "radiometer" system of comparable efficiency. After Sec. 3.2 it is left to the reader to make this substitution.

3.2 Koehler Illumination System

The usual requirement for an illuminating system is to produce a uniform illumination from a nonuniform light source, such as a lamp filament or an arc. *Koehler illumination* is the classical way of achieving this. As shown in Fig. 3.1, a *condenser* images the source in the pupil of the *projection lens*. The projection lens images the region of the condenser on the area to be illuminated. The illumination produced is the same as if a magnified source, located at the projection lens, were directly illuminating the target area. For a given projection lens diameter, the maximum illumination is produced when the lens pupil is filled by the image of the source. As indicated in Fig. 3.1, a simple spherical reflector centered on the source will image the source back on itself, increasing the average brightness of the source

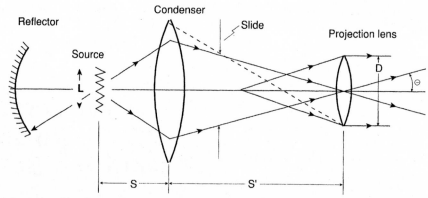

Figure 3.1 The Koehler illumination system is used in a projection condenser to produce even illumination from a nonuniform source. The condenser images the source in the pupil of the projection lens, and the projection lens images the slide, which is close to the condenser, on the screen. The screen illumination is then the same as if it were illuminated by an enlarged source placed at the projection lens. A spherical reflector with its center of curvature at the source images the source back on itself, filling in any gaps in the source and thereby increasing its average brightness.

and thus increasing the illumination on the target area. Condenser elements often have a molded aspheric surface; this can reduce or eliminate the spherical aberration, and also allows a higher-powered element for a given diameter.

To obtain high photometric efficiency, we might prefer the smallest possible source, so as to use the least power, generate the least heat, etc. The size of the smallest source whose image can fill the pupil of the projection lens is given by

$$L \geq \frac{D\Theta}{n} \tag{3.1}$$

where L is the linear source dimension, D is the diameter of the projection lens aperture, Θ is the half-field projection angle, and n is the index in which the source is immersed (almost always $n = 1.0$, for air). Reaching this limit requires that the condenser collect a full hemisphere of light from the source. This is quite difficult to achieve, and a value for L perhaps twice as large as given by Eq. (3.1) is more typical of an actual source size. (Incidentally, this limit applies to *all* optical systems, not just those analogous to the Koehler type.)

In radiometer usage the system is reversed; the illuminated area becomes the radiation source and the lamp becomes the detector. The system produces uniform illumination on the surface of the detector, and Eq. (3.1) indicates the smallest possible detector which will utilize the full aperture and field of the system. Note that for several classes of detector, e.g., lead sulfide, the limiting signal-to-noise ratio is inversely proportional to the size of the detector, i.e., the square root of the area, and a small detector is preferred for this reason. A Koehler system is often advantageous in a detector application because it will uniformly illuminate the surface of the detector; this alleviates the many problems which can arise because of a nonuniform response across the detector surface.

Sample calculations

A 35-mm slide projector has a 5-in focal length, $f/3.5$ projection lens. The slide diagonal is 1.7 in. What is the smallest filament which will fill the pupil through every point in the slide?

The projection lens pupil diameter is $D = 5$ in/3.5 = 1.429 in and the half-field angle is $\Theta = \frac{1}{2} \cdot 1.7$ in/5 in = 0.17. Thus the minimum filament size is

$$L = \frac{D\Theta}{n} = \frac{1.429 \times 0.17}{1.0} = 0.243 \text{ in}$$

Something closer to $\frac{1}{2}$ in would be a more realistic dimension.

Sample calculations

If the condenser in the previous calculation is located 1 in from the slide and the lamp filament is 0.5 in square, what is the (thin lens) focal length of the condenser, and what condenser diameter is required to fill the projection lens through the corners of the slide? With a 5-in focal length and a 1-in slide-to-condenser distance, S' in Fig. 3.1 is 6 in. The required condenser magnification is the lens pupil diameter divided by the source size, or $m = (-)1.429/0.5 = -2.857$, and by Eq. (1.6) $S = S'/m = 6$ in/$(-2.857) = -2.1$ in. Then the focal length is found using Eq. (1.4):

$$\frac{1}{6} = \frac{1}{f} + \frac{1}{-2.1}$$

$$\frac{1}{f} = 0.1667 + 0.4762 = 0.6429$$

$$f = 1.5556 \text{ in}$$

The minimum condenser diameter is defined by the ray (dashed in Fig. 3.1) from the bottom of the projection lens to the top (corner) of the slide. The ray slope is $\frac{1}{2}$(pupil + slide)/(focal length) = $\frac{1}{2}$(1.429 +

1.7)/5 = 0.3129. Projecting this ray to the plane of the condenser, the ray height on the condenser = (half the slide diagonal)+(ray slope)×(slide to condenser distance) = 0.85 + 0.3129×1.0 = 1.163 in and the required condenser diameter is 2.326 in. Note that the condenser could be made rectangular, in which case 2.326 in would be its diagonal.

Note that in these discussions we have tacitly assumed that the maximum illumination and a pupil filled with light were desired. As pointed out in Chap. 4, there are applications in which the pupil is deliberately not filled with light, or where the light is deliberately decentered in the pupil. This is based on diffraction considerations and is done to enhance the imagery.

3.3 Critical Illumination

What is usually called "critical illumination" is simply the illumination produced by focusing a source directly on the area to be illuminated. Unless the source is uniformly bright, uniform illumination cannot be obtained efficiently. An example of this type of illumination is the arc motion picture projector sketched in Fig. 3.2. A small arc

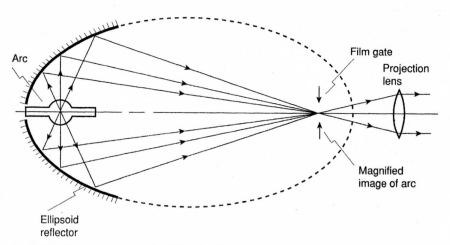

Figure 3.2 An example of critical illumination, where the source is imaged directly on the surface to be illuminated. An ellipsoidal reflector forms an image of a very nonuniform arc on the film of a movie projector. Because of the coma flare in the image and the multiple viewpoints from which the mirror "sees" the arc, the image is blurred out enough to produce acceptably uniform illumination. This system works only because the numerical aperture is large; with a small NA an image of the source will be projected on the screen. Ordinarily, critical illumination requires a uniformly bright source.

located at one focus of the ellipsoidal reflector is imaged (without spherical aberration) at the other focus of the ellipse. Although the arc brightness is very nonuniform, the illumination at the film gate is acceptably uniform. This is because the ellipsoid has a large amount of coma which smears out the image, plus the fact that each incremental area of the mirror "sees" and images the arc from a different viewpoint. All the different and differently oriented images combine to smooth out the light. The success of this illumination smoothing depends on the system having a large numerical aperture (i.e., a fast, small f-number, projection lens) so that these images are effectively blurred out. Note that the longer path from the source to the edge of the mirror means that the edge of the *beam* is less intense than the center. This is a common problem when conic reflectors are used to collect a very large solid angle of light.

3.4   Illumination Smoothing Devices

The *light pipe,* or *integrator bar,* is a simple device which can be used to produce a uniformly illuminated area. As shown in Fig. 3.3, when a nonuniform source is imaged on one end of the bar, the other end is illuminated by the checkerboard array of images which have been reflected from the walls. The number of reflected images is determined by the convergence angle of the light from the condenser and the length of the pipe. The light pipe cross section may be square, rectangular, or round. The pipe may be a solid bar or a hollow mirror array. If the pipe is tapered, the illuminated area can be larger or

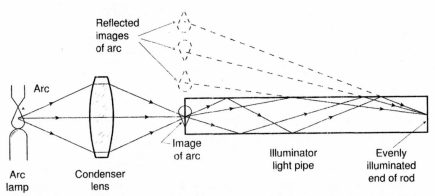

Figure 3.3 A light pipe can produce very uniform illumination from a nonuniform source. As shown here, the reflecting walls of the pipe form multiple images of the source. This array of images illuminates the output end of the pipe quite evenly, just as a similar array of real sources would.

smaller than the input end. Note that a tapered pipe changes the beam spread angle so that the spread angle at each end is approximately inversely proportional to the size of the pipe end faces.

Projectors which use arc lamps such as mercury, xenon, or the newer metal halide arcs often collect as much of the light as possible from the lamp by using a very deep aspheric reflector such as an ellipsoid or paraboloid. Because of the variation of lamp-to-mirror distance over the reflector surface, the outer parts of the beam are less intense than the inner parts of the beam (this is coma). In addition, the center part of the beam is obstructed by the arc envelope. Figure 3.4 shows a *beam homogenizer* array which can be used to smooth out the illumination. The light from the arc is collimated (imaged at infinity) by a deep paraboloid reflector. Each lens of the first array forms an image of the arc in the corresponding lens of the second array. The lens of the second array images the lens of the first array at infinity. Note that this combination behaves like the condenser and projection lens of a Koehler illumination system. All of these images are then focused on the film gate by the first condenser, where the combined images produce quite uniform illumination. The second condenser images the first condenser in the projection lens pupil. Again, the second condenser is working like a Koehler system condenser, imaging the nonuniform brightness first condenser into the projection lens.

3.5 Liquid Crystal Display (LCD) and Digital Micromirror Device (DMD) Projectors

A liquid crystal display (LCD) panel is often used in the projection of computer graphics. Two characteristics of the LCD must be taken into account in designing the illumination system. The first is that the contrast of the display is best when the illumination at the panel is collimated and its direction is approximately normal to the panel. (The optimum may actually be a few degrees from the normal in a specific direction.) A common way of accommodating this characteristic is shown in Fig. 3.5, where the light source is collimated (either by a refractive condenser as shown, or by a reflective system) and a convergent field lens is placed between the LCD and the projector lens to image the light source in the pupil of the projection lens. As shown, the field lens is often a fresnel lens, because the fresnel does not introduce any Petzval field curvature, as would an ordinary lens. To avoid having the fresnel grooves appear in the projected image, they are made quite fine and the fresnel is located well away from the plane of the LCD. This is acceptable because the system resolution, limited by the LCD pixel size, is well below the diffraction limit.

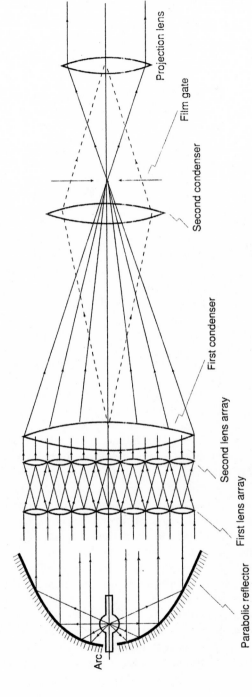

Figure 3.4 The beam homogenizer array of lenses is used to even out the intensity of the beam from the deep conic section reflector (in this case a paraboloid). A conic reflector is free of spherical but is afflicted with coma, which causes the beam intensity to fall off rapidly toward the beam edge. The first lens array images the source in the second array, which then images the lenses of the first array at the film gate. Each image is of different intensity, but they are all uniform and they all combine to produce efficient, even illumination.

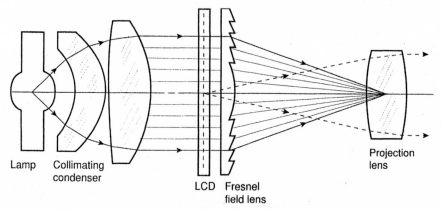

Figure 3.5 To achieve high contrast, a liquid crystal display (LCD) panel should be illuminated with collimated light. A fresnel lens, placed between the LCD and the projection lens, can be used as a condenser to focus the light into the pupil of the lens. The fresnel is used, despite its inefficiency and ring structure, because it does not cause the field curvature that an ordinary lens would introduce.

The second LCD characteristic affecting performance is the so-called aperture ratio of the LCD. This is the ratio of the transmitting area of the panel to the total area. This is always less than unity because, as shown in Fig. 3.6, the circuitry and transistors mask part of the area. For a three-color LCD, because each color (RGB) pixel occupies only one-third of the transmitting area, there is another loss factor (of three). An array of *microlenses* can converge the light (which would otherwise fall on the masked area) so that the light is condensed enough to pass through the transmitting area. Of course, this increases the convergence/divergence of the light beam so that a larger projection lens aperture is needed to pass the light. A second layer of microlenses on the output side of the LCD can be used to reduce the divergence.

It has been proposed to improve the three-color aperture ratio by directing the illuminating light in a different direction for each color, so that one large microlens can be used for each set of three pixels, and each color is obliquely directed to the appropriate color pixel. This can be accomplished by inserting tilted dichroic mirrors in the illuminating beam to produce a separate oblique beam for each color. Another approach has been to use diffractive microlenses so designed that each of three layers of diffractive microlenses is efficient only for one of the three colors; each lens covers the area of three color pixels but affects only its own color light to direct and focus it onto the proper color pixel. There are questions of bandwidth and diffraction efficiency which affect the viability of this scheme.

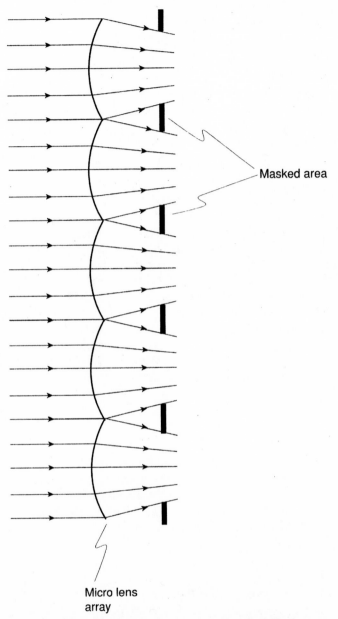

Masked area

Micro lens
array

Figure 3.6 The transmission of an LCD panel is partially obstructed by the masks and electronic structure defining each pixel. The
transmission efficiency can be improved by the use of microlenses,
one per pixel, which condense the illuminating beam so that it can
pass through the restricted aperture of the pixel.

A different approach than the LCD projector involves the use of what is called a *digital micromirror device,* or DMD. This is an array of pixel-sized mirrors, each flexibly mounted. The mirrors can be individually tilted (along their diagonal) so that the reflected light is caused to pass through or to miss the entrance pupil of the projection lens, thus producing a picture. The illumination for such a system must be oblique and is often critical illumination, usually modified in order to get even illumination on the screen. Color is produced by a rotating filter wheel; the gray scale is produced by extremely rapid oscillation of the mirrors to vary the on-off ratio.

3.6 Photometry, Radiometry, Illumination, Etc.

In order to simplify and compress this subject we treat radiometry with something less than great rigor in this section. For the vast majority of applications the effect of our simplifications will be negligible. But, for theoretical or extremely exacting work, the reader should refer to texts dealing with the subject in greater detail.

The *brightness* ($\equiv$ radiance or luminance) of a source is measured in watts (or lumens) output per solid angle per unit area, where the area is measured in a plane normal to the direction of view. The brightness of a perfect diffuser or of a self-luminous object is constant regardless of the direction of view. This is expressed as *Lambert's law,* which says that the *intensity* (in watts or lumens per solid angle) of a small area in the surface varies as

$$I(\Theta) = I_o \cos \Theta \tag{3.2}$$

where $I(\Theta)$ is the intensity in the direction Θ, I_o is the intensity in a direction normal to the surface, and Θ is the angle between the normal and the direction of view as schematically shown in Fig. 3.7A.

The relation between the total power emitted into a hemisphere and the brightness of the diffuse emitting surface is

$$B = \frac{P}{\pi} \tag{3.3}$$

where B is the surface brightness and P is the power emitted per unit area from the surface.

In any optical system, brightness is conserved. That is, if we neglect transmission losses, the brightness of an image is the same as that of the object, or

$$B_i = t \cdot B_o \tag{3.4}$$

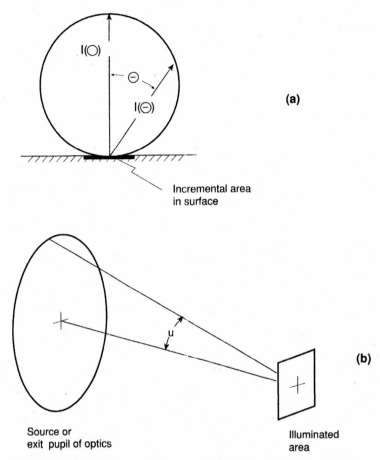

Figure 3.7 (A) According to Lambert's law, the intensity (power emitted per solid angle) of a small area in a diffuse surface varies with the cosine of Θ. (B) The illumination (incident power per unit area) produced on a surface is a function of the solid angle subtended by the illumination source, or subtended by the exit pupil of the imaging optics, as seen from the illuminated point.

Stated in another way, the brightness of an image cannot exceed the brightness of the object.

The illumination or irradiance (in watts or lumens per unit area) produced on a surface is given by

$$E = t \cdot B \cdot \Omega \cdot \cos \Theta \tag{3.5}$$

where t is the transmission factor, B is the source brightness, Θ is the angle at which the light is incident on the surface, and Ω is either (1)

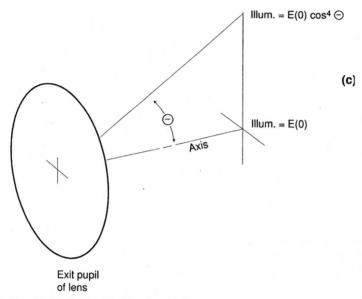

Illum. = E(0) cos⁴ ⊖

(c)

Illum. = E(0)

Axis

Exit pupil
of lens

Figure 3.7 (C) The illumination in the image on a plane varies as the
fourth power of the cosine of the angle of obliquity.

the solid angle subtended by the source as seen from the surface or
(2) the solid angle of the illuminating cone forming the image.

Note that the equivalence of 1 and 2 indicates that the exit pupil of
an optical system takes on the brightness of the source when viewed
from the image. For modest angles, the solid angle of the illuminating
cone can be found from

$$\Omega = \pi \cdot NA^2 \tag{3.6}$$

where NA is the numerical aperture, and $NA = n \cdot \sin u$, where u is
the half angle of the illuminating cone as illustrated in Fig. 3.7B, or
$NA = 1/(2 \cdot f\text{-}number)$.

The *cosine fourth* effect is a good approximation to the way that
illumination varies across the image plane of most optical systems.
When we consider how the solid angle subtended by the exit pupil of
a lens [Ω in Eq. (3.5)] varies across the flat image plane of a lens as
shown in Fig. 3.7C, we find that

$$\Omega(\Theta) = \Omega(O) \cdot \cos^3 \Theta \tag{3.7}$$

and substituting into Eq. (3.5), we get

$$E(\Theta) = E(O) \cdot \cos^4 \Theta \tag{3.8}$$

where $\Omega(O)$ and $E(O)$ are the solid angle and the illumination, respectively, at the axial image point, and $\Omega(\Theta)$ and $E(\Theta)$ are the solid angle and illumination at an angular displacement of Θ from the axis. Note that this reduction in the illumination is in addition to that caused by vignetting (Sec. 2.1).

The preceding relationships allow the calculation of any illumination project. The units which are commonly used in photometry are summarized below.

Intensity

candle (candela) $\equiv$ one lumen per steradian emitted from a point source (one-sixtieth of the intensity of one square centimeter of a blackbody at 2042 K)

Illumination

footcandle $\equiv$ one lumen per square foot incident on a surface

phot $\equiv$ one lumen per square centimeter

lux (meter candle) $\equiv$ one lumen per square meter

Brightness (luminance)

stilb* $\equiv$ one candle per square centimeter

$\equiv$ one lumen per steradian per square centimeter

lambert* $\equiv$ $(1/\pi)$ candles per square centimeter

foot-lambert* $\equiv$ $(1/\pi)$ candles per square foot

Sample calculations

How many lumens will a 35-mm slide projector produce? Assume a slide aperture 24 by 36 mm, a 5-inch (127-mm) focal length, $f/3.5$ projection lens, and a lamp with an average filament brightness of 1000 candles per cm². A two-element condenser with uncoated surfaces will have reflection losses of about $(1-0.96^4) = 15$ percent, so let's assume a condenser transmission of about 85 percent. A three-element projec-

---

*Note that the area in the brightness units is measured in a plane normal to the direction of view, *not* in the plane of the surface. This, combined with Lambert's law, indicates that a diffuse emitter has a constant brightness in all directions.

tion lens, if quarter-wave low reflection coated, should transmit about $0.993^6 = 96$ percent, so we have a system transmission factor of $0.85 \cdot 0.96 = 0.81$.

The pupil of the projection lens has a diameter of $5/3.5 = 1.429$ in $= 36.3$ mm, and an area of 1034 mm^2 or 10.34 cm^2. If we fill the pupil with the filament image, the pupil brightness will be $0.81 \cdot 1000 = 810$ candles per cm^2, and the pupil (if considered as a "small" source) will have an intensity of $810 \cdot 10.34 = 8376$ candles, or 8376 lumens per steradian.

If the 35-mm slide is projected to infinity (or to a "long" distance), the solid angle subtended by the image will be the slide area divided by the square of the projection lens focal length, or $24 \cdot 36/127^2 = 0.0536$ steradians. Thus the flux in the image will be $8376 \cdot 0.0536 = 449$ lumens.

Another approach might assume that the slide is to be projected to an image size of 4 by 6 ft. This is a magnification of $(4 \cdot 12 \cdot 25.4)/24 = -50.8\times$, and using the equations of Sec. 1.3, the image distance necessary to produce this magnification with a 5 in focal length lens is $s' = f(1-m) = 127(1 + 50.8) = 6579$ mm (or about 21 ft 7 in). Thus the lens pupil will subtend (from the screen) a solid angle given by its area divided by the square of this distance, or $1034/6579^2 = 2.39 \cdot 10^{-5}$ steradians. Using the pupil brightness of 810 candles per cm^2 (or 810 lumens per ster per cm^2), Eq. (3.5) gives us a screen illumination of

$$E = tB\Omega \cos \Theta = 810 \cdot 2.39 \cdot 10^{-5} \cdot 1.0 = 0.0194 \text{ lumens per cm}^2$$

At 929 cm^2 per ft^2, this is 18 lumens per square foot, or 18 footcandles illumination.

The area of the screen image is $4 \times 6 = 24$ ft^2, so that the lumen output is $24 \times 18 = 432$ lumens. The small difference between this value and the 449 lumens calculated above results from the different assumptions regarding the projection distance.

Sample calculations

In the preceding system, how large must the lamp filament be in order to fill the pupil of the projection lens from every point in the 35-mm slide? The minimum size is given by Eq. (3.1) as

$$L \geq \frac{D\Theta}{n}$$

where D is the projection lens diameter ($D = 36.3$ mm), Θ is the half-field projection angle, and n is the index in which the filament is immersed ($n = 1.0$). The projection angle is the slide diagonal (43.2 mm) divided by the projection lens focal length (or its object distance), and the half angle $\Theta = 21.6/127 = 0.17$, and we get

$$L \geq \frac{36.3 \cdot 0.17}{1.0} = 6.2 \text{ mm}$$

A reasonable, easily attained value might be about twice this minimum, or about 12 mm.

Sample calculations

With no slide in the projector, what will the screen brightness be? Assume that the screen is diffuse, with a 90 percent reflectance.

For a diffuse screen, the brightness can be easily calculated in footlamberts, by multiplying the illumination in footcandles by the reflectance, to get $18 \cdot 0.9 = 16.2$ foot-lamberts. Another approach uses the screen illumination of 0.0194 lumen per cm^2 times the 0.9 reflectance to get a total reflected flux of $0.9 \cdot 0.0194 = 0.0175$ lumen per cm^2. Dividing by π [per Eq. (3.3)], we get the screen brightness of

$$B = \frac{P}{\pi} = \frac{0.0175}{\pi} = 0.0056 \text{ lumen per ster per cm}^2$$

$$= 0.0056 \text{ candle per cm}^2$$

$$= 0.0056 \text{ stilb}$$

This assumes a diffuse or lambertian surface screen. Many projection screens use nonlambertian surfaces to increase the apparent brightness. The *screen gain* is the factor by which the brightness is increased over that of a perfect diffuser. Thus a screen with a gain of 2.5 would produce a brightness of $2.5 \cdot 0.0056 = 0.014$ stilb.

4

System Limits: Performance and Configuration

4.1 Introduction

It is the intent of this chapter to briefly outline some limits to which all optical systems must conform. These are: (1) limits of performance or resolution, (2) limits on throughput, and (3) limits on the relationships between beam angles and sizes. In the initial stages of system layout it is essential that these limits be known and harmonized with what is expected of the optical system. Occasionally this first step simply proves that "it can't be done." But even a negative result like this can be worthwhile if it avoids a waste of time spent on the physically impossible.

4.2 The Diffraction Limit

The image of a point source, formed by a perfect optical system with a uniformly transmitting circular aperture, is a diffraction pattern which consists of a circular central bright patch, called the *Airy disk,* surrounded by alternating dark and light concentric rings. The Airy disk contains 84 percent of the energy in the image. The first dark ring has a diameter given by

$$D = \frac{1.22\lambda}{\text{NA}} \tag{4.1}$$

where λ is the wavelength and NA is the numerical aperture of the imaging cone of light (NA $= n \cdot \sin u$). The peak illumination in the first bright ring is only 1.7 percent of the peak illumination in the Airy disk, and the illumination level in the other rings falls off quite rapidly. Figures 4.1 and 4.2 describe the diffraction pattern.

Point resolution is the ability to distinguish the images of two adjacent, closely spaced points. Obviously the diffraction blur affects this ability.

The *Rayleigh criterion* assumes that two points can be clearly resolved if they are separated by the radius of the first dark ring of the diffraction pattern, or

$$\text{Rayleigh separation} = \frac{0.61\lambda}{\text{NA}} \qquad (4.2)$$

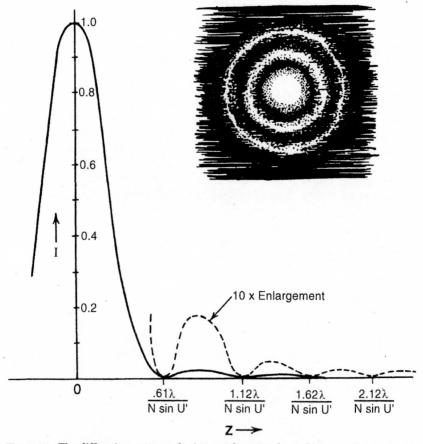

Figure 4.1 The diffraction pattern: the image of a point formed by a perfect lens with a circular aperture. The pattern consists of a bright central circular patch, called the Airy disk, surrounded by concentric rings of rapidly diminishing intensity.

Ring	Circular aperture		
	Z	Peak illumination	Energy in ring
Central maximum	0	1.0	83.9%
1st dark ring	0.61 λ/N' sin U'	0.0	
1st bright ring	0.82 λ/N' sin U'	0.017	7.1%
2d dark ring	1.12 λ/N' sin U'	0.0	.
2d bright ring	1.33 λ/N' sin U'	0.0041	2.8%
3rd dark ring	1.62 λ/N' sin U'	0.0	
3rd bright ring	1.85 λ/N' sin U'	0.0016	1.5%
4th dark ring	2.12 λ/N' sin U'	0.0	
4th bright ring	2.36 λ/N' sin U'	0.00078	1.0%
5th dark ring	2.62 λ/N' sin U'		

Figure 4.2 Tabulation of the characteristics of the diffraction pattern. Z is the radial dimension of the ring. Note that most (84 percent) of the energy is in the central patch, contained within the first dark ring, and the peak illumination in the first bright ring is only 1.7 percent of illumination at the center of the Airy disk.

The *Sparrow criterion* postulates a limit of about a 20 percent smaller separation, or

$$\text{Sparrow separation} = \frac{0.50\lambda}{\text{NA}} \qquad (4.3)$$

The *Dawes criterion* is an empirical one, derived from observations with the human eye as the sensor; it is only about 2 percent larger than the Sparrow criterion.

The resolution limit at the image is related to the resolution at the *object* by the magnification of the optical system. Since the magnification is equal to u/u' or sin u/sin u', one can determine the resolution limit at the object simply by using the object-side NA in Eq. (4.2) or (4.3). If the object is at infinity, the resolution must be expressed angularly, and the separation corresponding to the Rayleigh criterion is

$$\text{Rayleigh angular separation} = \frac{1.22\lambda}{P} \qquad (4.4)$$

where P is the diameter of the entrance pupil of the system. A commonly used version of Eq. (4.4) assumes a wavelength of 0.55 μm (the peak of the human visual response), that P is in inches, and that the angular resolution is in seconds of arc.

$$\text{Rayleigh angular separation} = \frac{5.5}{P} \qquad (4.5)$$

For the Sparrow criterion, the constant in Eq. (4.5) becomes 4.5; for Dawes it is 4.6.

Line resolution is the ability to separate or recognize the elements of a pattern of alternating high and low brightness parallel lines. An optical system is a low-pass filter, in that it cannot transmit information at a spatial frequency higher than the *cutoff frequency,* given (in cycles per unit length) by

$$v_o = \frac{2NA}{\lambda} = \frac{1}{\lambda \cdot f\text{-number}} \qquad (4.6)$$

This frequency corresponds to a line spacing equal to the Sparrow criterion in Eq. (4.3). It is an absolute cutoff, with zero contrast between the light and dark lines in the image. At a frequency corresponding to the Rayleigh criterion, the perfect lens image pattern has about 10 percent modulation [defined by Eq. (4.8) below]. In collimated space, i.e., with the object or image at infinity, the frequency is defined in cycles per radian, and the cutoff frequency is

$$\text{Angular } v_o = \frac{P}{\lambda} \qquad (4.7)$$

where P is the diameter of the entrance or exit pupil.

The *modulation transfer function* (MTF) describes the way that the optical system transfers contrast or modulation from object to image, as a function of spatial frequency. The *modulation* is defined as

$$M = \frac{\text{max} - \text{min}}{\text{max} + \text{min}} \qquad (4.8)$$

where max and min are, respectively, the maximum and minimum values of brightness (in the object) or illumination (in the image), and the object is a pattern of parallel lines whose brightness varies according to a sine function. The modulation transfer factor for a specific frequency is the ratio of the modulation in the image to that in the object, or

$$\text{MTF} = \frac{M_i}{M_o} \qquad (4.9)$$

For a perfect optical system the modulation transfer function is given by

$$MTF(v) = \frac{2}{\pi}(\phi - \cos\phi \sin\phi) \qquad (4.10)$$

and ϕ is defined as

$$\phi = \arccos\left(\frac{\lambda v}{2NA}\right) \qquad (4.11)$$

where v is the spatial frequency, λ is the wavelength, and $NA = n \sin u$ is the numerical aperture. Note that the term within parentheses in Eq. (4.11), being a cosine, cannot exceed unity; this then is the source of Eq. (4.6) for the cutoff frequency v_o.

Figure 4.3 shows the MTF plotted against spatial frequency (which has been normalized to unity for the cutoff frequency) for a perfect optical system (curve A), as well as for a perfect system defocused by various amounts. Note that curve B results from an amount of defocus which produces a wavefront deformation, or OPD, equal to a quarter wavelength ($\lambda/4$). This is the *Rayleigh limit* for a tolerable amount of aberration, yielding an image which is "sensibly perfect." An optical system with this amount (OPD = $\lambda/4$) of wavefront aberration is often described as *diffraction limited* (although it obviously is not).

When the pupil of the optical system is *not* uniformly illuminated (as assumed above) the diffraction pattern will differ from that described in Figs. 4.1 and 4.2. For example, if the wavefront intensity varies as a gaussian or exponential (as in a laser beam), the diffraction pattern is also a gaussian distribution.

The center of the pupil transmits the low spatial frequency information and the outer portions transmit the high spatial frequency information. In some optical systems where a high image modulation is required for the low spatial frequencies and a low modulation is acceptable for the high frequencies (as in microlithography) the illumination (condenser) system is designed so that only the center of the pupil is filled with light. Called semicoherent illumination, this produces very high contrast images at low frequencies (at the expense of the high). Decentering the illuminated portion of the pupil can emphasize the contrast of features which have a strongly directional characteristic.

4.3   Image Sensor Limits

Any sensor has a performance limit. The eye, film, an image tube, a CCD, etc., all have performance or resolution limits imposed by their structure and operating characteristics. These limits can be described in detail by MTF, or by what is even more useful, a threshold curve.

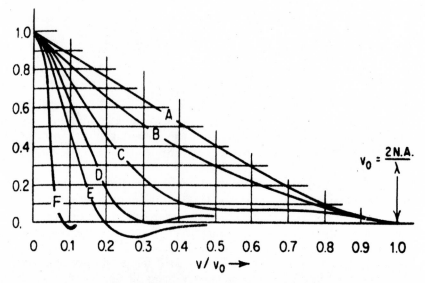

The effect of defocusing on the modulation transfer function of an aberration-free system.

(a) In focus	OPD = 0.0
(b) Defocus = $\lambda/2N \sin^2 U$	OPD = $\lambda/4$
(c) Defocus = $\lambda/N \sin^2 U$	OPD = $\lambda/2$
(d) Defocus = $3\lambda/2N \sin^2 U$	OPD = $3\lambda/4$
(e) Defocus = $2\lambda/N \sin^2 U$	OPD = λ
(f) Defocus = $4\lambda/N \sin^2 U$	OPD = 2λ

(Curves are based on diffraction effects—not on a geometric calculation.)

Figure 4.3 The modulation transfer function (MTF) of a perfect lens for various amounts of defocusing. The image contrast indicated by curve A cannot be exceeded for an ordinary lens. Curve B indicates the MTF for a system with a wavefront deformation (OPD) equal to a quarter of the wavelength of light, corresponding to what is called the "Rayleigh criterion." A system with this level of performance is often (inaccurately) called "diffraction limited."

This curve describes, as a function of spatial frequency, the minimum image modulation necessary to produce a response (i.e., to recognize a line pattern in the image). A plot of the threshold modulation against frequency is often called an *AIM curve* (for aerial image modulation). When plotted on the same graph as in Fig. 4.5, the intersection of the image modulation (MTF) curve and the sensor AIM curve indicates

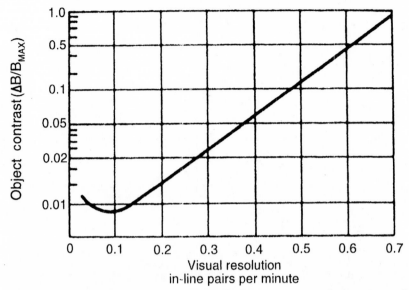

Figure 4.4 The object contrast necessary for the human eye to resolve a pattern of alternating bright and dark bars of equal width. Note that this curve shifts upward for lower light levels and drops at higher levels. For this plot the bright bars had a brightness of 23 foot-lamberts.

the limiting resolution frequency. Note that the AIM curve may vary with conditions such as the illumination level or exposure. Figure 4.4 shows a "contrast" threshold curve for the human eye at a particular level of target brightness. As one might expect, the finer the detail, the higher the contrast required to resolve it. This curve moves down with higher target brightness and up with lower.

Often a single number "resolution" is useful, despite the fact that, as Fig. 4.5 shows, resolution is but a single value attempting to represent what is a rather complex relationship. Thus one minute of arc is, for better or worse, often used as the performance limit for the eye, one over twice the pixel spacing is used for a CCD or the like, and measured resolution values are available for various types of film.

4.4 Diffraction Limit vs. Sensor Limit

It is important to be sure that the performance limits due to diffraction, and those imposed by the sensor of your system have a relationship that produces the kind of results which are best for your application. As an example, let's consider a visual telescope with a 1-in-diameter objective lens (and entrance pupil). The Rayleigh resolution limit according to Eq. (4.5) is 5.5 seconds of arc in object space. If our telescope has a

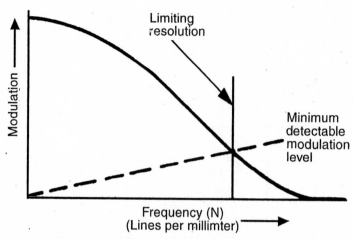

Figure 4.5 When the MTF curve of the optics and the AIM curve (the minimum detectable modulation) of the sensor are plotted together, the intersection indicates the maximum spatial frequency at which the line pattern can be detected. This is usually called the resolution of the system.

magnification of 11×, this 5.5 seconds is magnified and presented to the eye as 11×5.5 = 60 seconds = 1 minute of arc, which we presume to be the resolution limit of the observer's eye. Now if we increase the magnification of the telescope, the image will be made larger, but, because of the diffraction at the objective lens, there will be no more information in the image than there was at 11× magnification. The increased magnification is called *empty magnification* for this reason. However, the added magnification is not totally useless, because (within limits) the larger image makes it easier for the observer to recognize a target and to point the telescope accurately. Consider a surveyor's transit with an objective diameter of 1 in, or 1¼ in, which may have a power of 25× or 30×, about twice the level at which empty magnification sets in. This extra power aids the surveyor in making accurate angle measurements. Many scopes have a significant amount of so-called empty magnification, and often a good case can be made to exceed the limits of Eqs. (4.12) and (4.13) (below) by a factor of several.

Now let's consider a riflescope. Here again, the objective diameter is often to the order of 1 in, but the magnification is usually low, in the range of 2× to perhaps 5×. There is obviously detail in the image which the eye cannot see because of the small size of the detail. The scope power is kept low in order to provide a large exit pupil and wide field, so that hunters can quickly bring the rifle to their shoulders and center their eyes in the exit pupil of the scope. The extra detail in the image provides, as a bonus, a crispness to the image which aids in the rapid recognition of objects in the field of view (a valuable bonus in light of the potential lethality of a mistaken discharge of the rifle).

We can express the "empty magnification" threshold for visual telescopes and afocal systems as follows:

$$MP(max) = 11 \cdot P \qquad (4.12)$$

where P is the entrance pupil diameter in inches, MP(max) is the magnification at which empty magnification begins, and we assume a visual resolution of 1 minute of arc. A similar limit for microscopes is

$$MP(max) = 218 \cdot NA \qquad (4.13)$$

where NA is the numerical aperture (NA = $n \sin u$) of the microscope objective. It is not uncommon for these "limits" to be exceeded by factors of several, and it has been suggested that under certain conditions, a magnification of 5 to even 10 times that indicated in Eqs. (4.12) and (4.13) may be optimum for some tasks.

4.5   The Optical Invariant

The invariant, given in Sec. 1.8 as INV = $n(y_p u - y u_p)$ can also be written as

$$INV = n(y_1 u_2 - y_2 u_1) \qquad (4.14)$$

where y_1, y_2, u_1, and u_2 are the paraxial ray heights and ray slopes of any two "different" rays (i.e., rays which have different axial intersection points and are therefore not scaled versions of each other). INV has the same value everywhere in the optical system. Two applications of the invariant are particularly useful. One was discussed in Sec. 1.8, where Eq. (1.22), which evaluated the invariant for axial and principal rays at the object and image planes, showed that

$$m = \frac{h'}{h} = \frac{nu}{n'u'} \qquad (4.15)$$

which means that, in air, the slope of the axial ray at the object *must* equal the ray slope at the image times the system magnification.

Similarly, if the invariant is evaluated at the entrance and exit pupils of a system (where the principal ray height y_p is zero), we get INV = $nyu_p = n'y'u_p'$, where y and y' are the axial ray heights at the entrance and exit pupils and u_p and u_p' are the half-field angles. This leads to our expression for the angular magnification of an afocal,

$$\text{MP} = \frac{y}{y'} = \frac{n'u_p'}{nu_p} \tag{4.16}$$

This means that, if you use an afocal device to reduce the beam diameter, the field of view (or beam divergence) will be increased in proportion to the beam diameter reduction.

4.6 Source and Detector Size Limits

There is a limit on the smallness of a source for an illumination system and on the smallness of a detector in a radiometric application. That limit was expressed by Eq. (3.1) as

$$L \geq \frac{D \cdot \Theta}{n}$$

where L is the minimum size of either the source or detector, D is the diameter of the projection lens of an illuminator or the diameter of the radiometer objective, Θ is the half-field angle that the device covers, and n is the index of refraction of the medium in which the source or detector is immersed. This value of L is the smallest possible if the full aperture D and the full field $2 \cdot \Theta$ are to be utilized. L is minimal if the axial ray at the source or detector has a slope of 90° and $\sin u = 1.0$. This is exceptionally difficult to achieve, and a value of L which is about twice as large as this limit is often encountered in practice.

4.7 Depth of Focus

The basic idea behind the concept of depth of focus is that there is some limit on the smallness of the detail in the image that the system can record, and that the system may be defocused by an amount which causes a blur of that size without a degradation in the recorded image quality. In photography the size of the darkened silver grains in the emulsion presents just such a limit. Pixel size is often used as a basis for the limit in other applications. If we make the assumptions that the optical system is perfect and that diffraction effects are negligible, the blur created by defocusing is given by

$$B = \frac{\delta}{f\text{-number}} \tag{4.17a}$$

or by

$$B = 2 \cdot \delta \cdot \text{NA} \tag{4.17b}$$

where B is the diameter of the blur, δ is the distance that the image is out of focus, f-number is the relative aperture, and NA is the numerical aperture (NA $= n \sin u$). Thus, if B represents the diameter of the tolerable blur, the depth of focus is

$$\delta = \pm B \cdot (f\text{-number}) \qquad (4.17c)$$

or

$$\delta = \frac{\pm B}{2 \cdot NA} \qquad (4.17d)$$

The *depth of field* is the distance that the object must be away from the nominal position of focus to create the same size blur. Note that the depth of field is related to the depth of focus by the longitudinal magnification. The close focus distance is

$$D_{near} = \frac{fD(A + B)}{fA - DB} \qquad (4.18)$$

and the far distance is

$$D_{far} = \frac{fD(A - B)}{fA + DB} \qquad (4.19)$$

where D is the nominal focus distance (since D is measured to the left of the lens, it is normally a negative number by our sign convention), f is the system focal length, and A is the system aperture (pupil) diameter.

The *hyperfocal distance* is the closest distance at which the system may be focused and still have an object at infinity "in focus":

$$D_{hyp} = \frac{-f \cdot A}{B} \qquad (4.20)$$

The depth of field then extends to half the hyperfocal distance.

Note well that the assumption of a perfect image without aberrations and without diffraction is an incorrect one. Nonetheless, these equations are a reasonable enough model of the situation that they are widely used in photography and other applications.

The *Rayleigh limit* for defocusing is based on the quarter-wave criterion for wavefront deformation, as mentioned in Sec. 4.2 and illustrated in Fig. 4.3. The amount of defocusing which will produce a quarter wave ($\lambda/4$) of wavefront deformation is

$$\delta = \frac{\pm \lambda}{2 \cdot n \cdot \sin^2 u} \qquad (4.21a)$$

or

$$\delta = \pm 2\lambda \bullet (f\text{-number})^2 \qquad\qquad (4.21b)$$

Note that this is a completely different criterion than that described above for the photographic depth of focus.

5

How to
Lay Out a System

5.1 The Process

In this chapter we are concerned with the actual process of determining the component powers and spacings to produce a system which will satisfy the requirements of whatever application is at hand. Our solution will consist of a set of component powers (ϕ_a, ϕ_b, ϕ_c, etc.) and a set of spacings (D_a, D_b, etc.). It may also include aperture or field stops and the diameters of the components and stops. This is a "thin lens" solution, and it is a (necessary) preliminary to the lens design process.

The process outlined here is intended to be quite general and broadly applicable. As such, the outline surely will suffer from overinclusiveness, and it will probably be more extensive than any single project will require. My apologies for this, but I hope it will serve the reader as a checklist or perhaps even as a source of ideas.

The process can be broken down into four steps:

1. Define

2. Restate

3. Solve

4. Optimize

A discussion of each step follows:

1. **Define.** The process should always begin with the system definition, i.e., a listing of the required system characteristics and the tasks which the system is to perform. Appropriate items might include:

Physical size (length, diameter)

Spatial limits (clearances, windows, bends)

Image size

Focal length

Magnification

Image orientation

Image location

Field(s) of view

Aperture (pupil) size

Numerical aperture or f-number

Wavelength and bandpass

Radiometric requirements (illumination, brightness)

Illumination uniformity (vignetting)

Resolution or performance

Characteristics of the sensor

Obviously, many of these items are redundant and many will not be applicable to a particular problem. However, it is always wise to make a list of clearly defined and agreed upon system requirements at the initiation of any project; such a list can also serve as a useful record if the project direction is changed in midstream.

2. Restate. Having defined the system, the list is reduced to those requirements which are determined by the first-order characteristics of the system, such as image size, image location, and image orientation. The various limits discussed in Chap. 4 should be taken into consideration at this point in order to be certain that the defined system is in the domain of the possible (or better still, the practical). It is often useful for the first-order system requirements to be stated in terms of the effect that the system has on the path of certain rays. Often these rays are the axial marginal ray (from the foot of the object through the margin of the aperture stop) and the principal ray (from the edge of the field through the center of the stop).

3. Solve. The next step is to determine, or solve for, the set of powers and spaces. Often a system will consist of just two components. In this case the equations of Sec. 1.10 can provide the answer without any further ado. In many cases the system will be similar to one of the basic systems described in Chap. 2, or it may be made up of some combination of these systems. In these cases the solution effort may be minimal. Occasionally, however, the requirements are such that these standard approaches don't provide the full answer, and a novel approach is necessary. Three useful ways to attack this situation are discussed in Sec. 5.2.

4. Optimize. The final stage is to select the optimum configuration (if there is more than one candidate arrangement) and/or to optimize the system. Often one configuration will be better suited to the overall situation for some reason. Note that here we do not mean optimize in the sense of correcting or balancing aberrations as is done in the lens design process. But we do want the first-order layout which will be the best in terms of its *potential* as a finished lens design. In this direction the considerations are quite simple: the weaker the power of a component, the better. A low-power component has less aberration, costs less, and is less sensitive to fabrication and assembly errors such as misalignment. An easy approach is to try to minimize the sum of the component powers (or really, the sum of their *absolute* powers). There are many other ways to weight the component powers. If the necessary diameters have been determined, the product of power and diameter is a rough measure of the cost, complexity, weight, etc., of the component which may result from the lens design process. The "work" done in bending a ray (which is simply the product $y\phi$) is another measure of difficulty; an even distribution of "work" among the various components is often desirable. Note that, in general, a longer system will tend to have weaker components; thus, if you have the space, you get a better system by using it fully.

5.2   The Algebraic Approach

This is the most elegant, useful, and laborious way to approach the problem of solving for a layout which satisfies our requirements. Those who detest algebra would be well advised to avoid this. However, it does, by its nature, provide *all* possible solutions, whereas the numerical and computer techniques outlined in the next two sections usually find only the solution nearest to the form envisaged by the system designer. For example, the algebra may lead to a solution equation which includes a square root, indicating that there are two solutions, one of which may turn out to be an unexpected and delightful surprise. The drawback to the algebraic approach is, to be blunt, the algebra. The probability of error is quite high; for many of us the probability of an error will significantly exceed unity. For this reason, one should repeat the algebra until the identical result is independently obtained at least twice.

Basically the method involves tracing paraxial rays symbolically, that is, using symbols instead of numbers for the values of the powers and spacings of the system. The rays to be traced are those which define the first-order characteristics required of the system. For example, in an afocal or telescopic system one would trace a ray entering the system with a slope of zero ($u_a = 0.0$) and an arbitrarily chosen ray height (y_a). Then to make the system afocal, we would solve the equation for the final ray slope $u_k{}'$ so that it had a value of zero. Using the same symbolic raytrace, we would solve the equation for the final ray height y_k to make its value equal to the initial ray height divided by the desired magnification of the telescope ($y_k = y_a/\text{MP}$). If the telescope had a requirement for an eye relief (exit pupil distance), the ray to be traced would be the principal ray, passing through the center of the aperture stop. If the stop were at the first component, the starting ray data would be $y_a = 0.0$, with the slope u_a chosen to match the half-field of the system. Then the eye relief is the intersection length of this ray after the last lens, or $l'_k = -y_k/u_k{}'$. Requirements for system length or component locations can be expressed as equations in the form of $D_a + D_b + \text{etc.} =$ desired sum. Diameter requirements or limitations can be expressed as the sum of the absolute values of the (vignetted) axial ray height plus the principal ray height, i.e., ($|V \cdot y| + |y_p|$), where V is the vignetting factor. All of the first-order properties, as well as the spatial and length requirements, can be expressed as equations. A simultaneous solution then yields the system layout.

For a system of any complexity and with a number of requirements and constraints, this is obviously not a trivial task. It is entirely possible that you can wind up with a set of equations which is beyond your ability to solve. In this case your time has not been entirely wasted; the equations can become a convenient basis for a numerical approach to the solution.

In order to ease the pain of working through the symbolic raytrace, we provide the following equations. These are the completely general equations for the heights and slopes of a ray with the starting data of y_a and u_a. Equations are given for the ray after it has passed through one, two, three, and four components. (Four should be enough for *anybody!*) Note that, because of the "scaleability" of paraxial rays, either y_a or u_a may be set equal to 1, provided the other is adjusted to maintain the necessary axial intersection distance for the ray ($l_a = -y_a/u_a$). This will greatly simplify the expressions. As an extreme example of simplification, note that, for the axial ray from an object at infinity, we may set $u_a = 0.0$ and $y_a = 1.0$.

The equations are for a system of components with powers ϕ_a, ϕ_b, ϕ_c, ϕ_d which are spaced apart by distances of D_a, D_b, D_c, where, for example, D_a is the distance from component a to component b, D_b is the distance from b to c, etc.

Given. u_a and y_a

For one component (see Fig. 5.1a)

$$u_b = u_a{'} = u_a - y_a\phi_a \qquad (5.1)$$

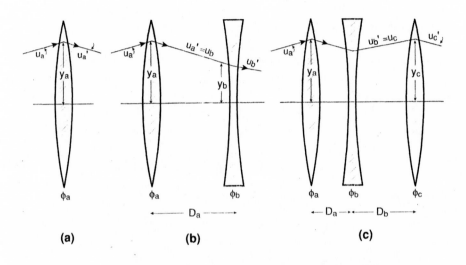

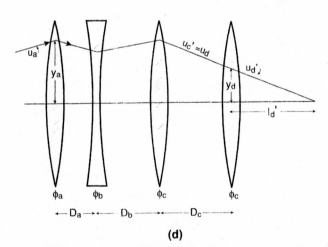

Figure 5.1 Schematic reference sketch for symbolic raytracing Eqs. (5.1) through (5.7). (A) Single component, Eq. (5.1). (B) Two components, Eqs. (5.2) and (5.3). (C) Three components, Eqs. (5.4) and (5.5). (D) Four components, Eqs. (5.6) and (5.7).

For two components (see Fig. 5.1*b*)

$$y_b = y_a + u_b D_a$$

$$= u_a D_a + y_a(1 - \phi_a D_a) \tag{5.2}$$

$$u_c = u_b{}' = u_b - y_b \phi_b$$

$$= u_a(1 - \phi_b D_a) - y_a(\phi_a + \phi_b - \phi_a \phi_b D_a) \tag{5.3}$$

For three components (see Fig. 5.1*c*)

$$y_c = y_b + u_c D_b$$

$$= u_a(D_a + D_b - \phi_b D_a D_b)$$

$$+ y_a[(1 - \phi_a D_a)(1 - \phi_b D_b) - \phi_a D_b] \tag{5.4}$$

$$u_d = u_c{}' = u_c - y_c \phi_c$$

$$= u_a[(1 - \phi_b D_a) - \phi_c(D_a + D_b - \phi_b D_a D_b)]$$

$$- y_a[\phi_a + \phi_b + \phi_c - \phi_a D_a(\phi_b + \phi_c)$$

$$- \phi_c D_b(\phi_a + \phi_b) + \phi_a \phi_b \phi_c D_a D_b] \tag{5.5}$$

For four components (see Fig. 5.1*d*)

$$y_d = y_c + u_d D_c$$

$$= u_a[D_a + D_b + D_c - \phi_b D_a(D_b + D_c)$$

$$- \phi_c D_c(D_a + D_b - \phi_b D_a D_b)]$$

$$+ y_a\{[(1 - \phi_a D_a)(1 - \phi_b D_b) - \phi_a D_b$$

$$- D_c[\phi_a + \phi_b + \phi_c + \phi_a \phi_b \phi_c D_a D_b$$

$$- \phi_a D_a(\phi_b + \phi_c) - \phi_c D_b(\phi_a + \phi_b)]\} \tag{5.6}$$

$$u_e = u_d{}' = u_d - y_d \phi_d$$

$$= u_a\{(1 - \phi_b D_a) - \phi_c(D_a + D_b - \phi_b D_a D_b)$$

$$- \phi_d[D_a + D_b + D_c - \phi_b D_a(D_b + D_c)$$

$$- \phi_c D_c(D_a + D_b - \phi_b D_a D_b)]\}$$

$$- y_a \{ \phi_a + \phi_b + \phi_c - \phi_a D_a (\phi_b + \phi_c)$$

$$- \phi_c D_b (\phi_a + \phi_b) + \phi_a \phi_b \phi_c D_a D_b$$

$$+ \phi_d (1 - \phi_a D_a)(1 - \phi_b D_b) - \phi_a \phi_d D_b$$

$$- \phi_d D_c [\phi_a + \phi_b + \phi_c + \phi_a \phi_b \phi_c D_a D_b$$

$$- \phi_a D_a (\phi_b + \phi_c) - \phi_c D_b (\phi_a + \phi_b)] \}} \tag{5.7}$$

5.3   The Numerical Solution Method

This approach might be called "cut and try, with differential adjustment." It is beloved by those who detest algebra. It works well provided that the type of system needed is reasonably well understood by the person doing the layout. And it is probably the most popular and widely used approach to optical system design. In essence one more or less arbitrarily selects one or two parameters (i.e., powers and/or spaces) and then adjusts the balance of the system to satisfy one or two of the system requirements. One parameter is then selected as a free variable. It is changed by a small increment, and the balance of the system is again adjusted. The change in an uncontrolled (but required) system characteristic is noted, and a differential solution for the free variable is made.

Sample calculations

As an excruciatingly trivial and simple example, let's lay out a keplerian telescope with a magnification of $-4\times$ and an overall length of 10 in. The system is sketched in Fig. 5.2. Of course, it doesn't take an Einstein to figure out the correct answer: $f_a = +8$ in and $f_b = +2$ in. But let's demonstrate the "cut, try, and adjust" process. We can,

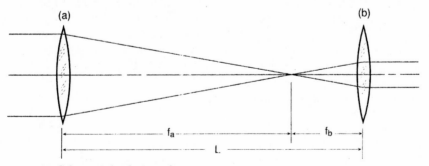

Figure 5.2   Schematic keplerian telescope.

with reference to Fig. 5.2, start by choosing f_a, then determine f_b to satisfy *either* the length or magnification requirement; let's go with magnification. Starting with a (deliberately bad) guess of $f_a = +5$ in, f_b must equal $-f_a/(-4) = +1.25$ in to get a magnification of $-4\times$. This gives us a length of $f_a + f_b = 6.25$ in. Next we choose $f_a = 6$ in and determine that f_b must be 1.5 in, giving us a length of 7.5 in. The length has changed $+1.25$ in for a $+1$-in change in f_a, so $(\delta L/\delta f_a) \approx 1.25/1.0 = 1.25$. We need a 10-in length, 2.5 in longer than our second trial; the solution is to increase f_a by $2.5/1.25 = 2$ in to a value of $f_a = +8$ in. Then f_b will be $+2$ in and the length will be 10 in, as required. Of course, all this is already worked out in Chap. 2 as Eqs. (2.4) and (2.5).

Sample calculations

In order to make this exercise a bit more interesting, let's make our telescope a lens-erecting type, again 10-in long, with a magnification of $+4\times$, and let's also require an eye relief of 4 in. This type of scope is sketched in Fig. 5.3. Again we play the same sort of game, arbitrarily selecting a focal length for the objective $f_a = +4$ in, and also a focal length for the eyelens $f_c = +1$ in. This pair has a magnification of $-4/1 = -4\times$, so we need a magnification of $m = -1.0\times$ from the erector to get a power of $+4\times$ for the total scope. The objective and eyelens focal lengths add up to a distance of $f_a + f_c = 4 + 1 = 5$, leaving a space of $10 - 5 = 5$ in for the erector, which at $m = -1.0$ must have $s = -2.5$ in and $s' = +2.5$ in. Equation (1.4) can be solved for the erector focal length $f_b = +1.25$ in. Now we can trace a principal ray through the center of the objective lens, using Eqs. (1.19) and (1.20) [or perhaps Eqs. (5.4) and (5.5)]. We find that this ray intersects the axis at 2.05 in from the eyelens; our eye relief is short by $(4 - 2.05) = 1.95$ in.

Next we try $f_a = +4$ (again) and $f_c = +1.25$ ($\delta f_c = +0.25$, or $\delta \phi_c = -0.2$). This pair has a power of MP $= -4/1.25 = -3.2$ and uses up a length of $4 + 1.25 = 5.25$ in, leaving 4.75 in for the erector to produce a magnification of $4/(-3.2) = -1.25\times$. Using the relationships $m = s'/s = -1.25$ and $s - s' = 4.75$, we find that $s = -2.111...$ and $s' = +2.63888...$, and again Eq. (1.4) gives us the erector focal length as $f_b = 1.172840$ in. Tracing the principal ray with Eqs. (1.19) and (1.20), we now calculate an eye relief of 2.565789; an increase of 0.515789 in the eye relief (ER) has been produced by an increase of eyelens focal length of 0.25 in.

So we have $\delta ER/\delta f_c \approx +0.515/+0.25 = +2.063154$ (or $\delta ER/\delta \phi_c \approx -2.578943$). Since we need an increase in the eye relief of $(4.0 - 2.565784) = +1.434211$ from our next try, we want a change in

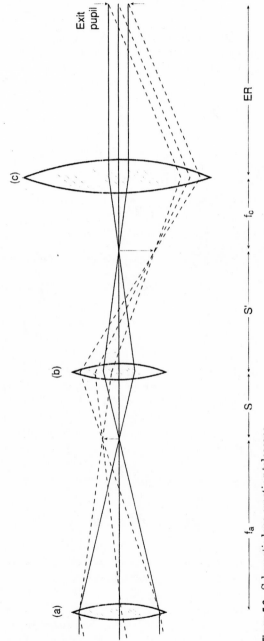

Figure 5.3 Schematic lens-erecting telescope.

the eyelens focal length of $+1.434/+2.063 = +0.695155$ to $f_c = 1.945155$. [Alternately, we could take a *power* change of $+1.434/(-2.578943) = -0.556123$ to get $\phi_c = +0.243876$ and $f_c = +4.100437$. There is a significant difference between the result we obtain using power compared to that using focal length. This is due to the nonlinearity of the relationships involved. Ordinarily using power works out better, but in this case your author's a priori knowledge indicates that the focal length result is the better one, so we will go with that.] Again, $f_a/f_c = 4$ in/1.945 in = 2.056391×; this leads to an erector magnification of $-4/2.056 = -1.945155×$ in a distance of $(10-4-1.945) = 4.054845$ in. The erector conjugates are $s = -1.376785$ in and $s' = +2.678060$ in, and $f_b = +0.909310$ in. The principal raytrace indicates an eye relief of $l_c' = +4.334309$ in. We're getting closer.

Our last change in f_c was $+0.695155$ in, and it produced a change in ER of $+1.768525$ in; thus we get $\delta ER/\delta f_c \approx 2.544073$. To change the ER by -0.334 in, we need a change in f_c of $-0.334/2.544 = -0.131407$ in, for a new f_c of $+1.813748$ in. This process can be carried on until the eye relief is close enough to the desired 4-in value. (The exact solution is $f_a = 4.0$ in, $f_b = +0.9529$ in, $f_c = 1.8298$ in.) Note that when the final value to be determined (the eye relief in this case) is one where an *exact* value is *not* required, the process can be terminated as soon as a reasonably close result is obtained.

In the above calculations we arbitrarily chose the focal length of the objective lens as $f_a = +4.0$ in. This is obviously an unused variable parameter, and we could, if desirable, use it to control an additional characteristic of the system. Thus we could repeat the above process for several additional values of f_a and choose from among the solutions the one that best suited our application. In this case we could, as indicated at the end of Sec. 5.1, select a solution on the basis of minimizing the component power sum.

5.4 First-Order Layout by Computer Code

An "automatic lens design" program can quickly and efficiently carry out the same process as outlined above. Every optical software program worth its salt can be set up with first-order targets in its merit function. What we are suggesting here is that the merit function be set up with *only* first-order targets, without any aberration or performance targets. These first-order targets would consist of the ray heights and ray slopes of the (paraxial) axial marginal ray and the heights and slopes of the (paraxial) principal ray. These targets can be chosen so as to control focal length, magnification, image sizes and locations, pupil size and location, beam diameter and slope, lens diameters, or almost anything you can think of. Many programs have some of these targets explicitly predefined. In addition to laying out zoom and multiconfiguration systems, a program's "multiconfiguration" (or "zoom") feature can be used to specify data for more than the two specific rays cited above (by using different aperture and field sizes and different stop positions for the alternate configurations).

In addition to the optimization routine, most programs have simple paraxial "solves" available. These will solve for a radius to produce a desired ray slope for either the axial or principal ray. They will also solve for a spacing to produce a desired ray height at the next surface, again for either the axial or principal ray. In the same vein, one can specify that two radii or two spaces must be equal or must differ by some amount. These program features are exact solutions of the paraxial raytrace equations, carried out without recourse to the optimization part of the program.

The optical system can be defined as a system of thin lens components, or if desired, thicknesses can usually be introduced without difficulty. The key is that only one surface of each component is used as a variable. In a thin lens system, each component is represented as a planoconvex or planoconcave element of zero thickness. The curvature of the curved surface is the variable, and the other surface is fixed. The airspaces between the elements are also variables.

The program optimization routine works in the same way as the numerical process outlined above. It is given a starting system, a set of targets, and a set of variables. It makes a small change in each variable (one at a time) and calculates the change produced in each target value, creating a matrix of the partial derivatives of the targets with respect to the variables. It assumes that the relationships involved are linear and makes a damped least squares (DLS) solution. It iterates this process until a solution is arrived at, or until no further progress is made toward a solution.

Note well that we are targeting *only* the first-order properties of the system. We deliberately omit any target which is related to image quality. This is, in general, a wise move even in the initial stages of a lens design process. The reason for this is that, if one targets aberrations before the first-order properties are close to their proper values, the software may find a "local optimum" which has high image quality but is far from satisfying the first-order requirements unless the merit function is heavily weighted in favor of the first-order targets.

If the design problem is underconstrained, that is, if there are more variable parameters than targets, the program will find the solution nearest to the starting system; it is designed to go for the solution with the minimum change. It is apparent that some "feel" for the type of system that is under consideration is a big help in choosing a starting form. One can buy some "feel" quite easily by trying several different starting systems and studying the results.

If you ask the program to solve an overconstrained problem (where there are more targets than variables), the odds are that you won't get an exact solution for all the targets. The DLS solution will minimize the sum of the squares of the errors (the differences between the target values in the merit function and the actual values of its solution). You can make some adjustments by juggling the weighting of the targets in the merit function to better control the important targets (at the expense of the others).

Many software programs have both "minimized" and "constrained" targets in the merit function. The program attempts to make the "constrained" targets go to the exact value specified, as opposed to minimizing the rms value of the errors for the "minimized" targets. If one were to do the lens-erecting telescope of the second example in Sec. 5.3 on a computer, the values for length, magnification, afocality, and eye relief might be set up as constraint targets and the sum of the component powers as a minimize variable.

5.5 A Quick Rough Sketch

When the power and space layout has been completed, a rough sketch, drawn to scale, is a worthwhile investment of your time. The components can be drawn as single lens elements to start, either as planoconvex or as equiconvex shapes (or planoconcave or equiconcave). The process is simple because the radius of a planoconvex or planoconcave lens is equal to the focal length multiplied by $(n-1)$. Thus for a glass lens with an index of 1.5, the radius is just half the focal length. If the index is higher, the radius is longer; for an index of 2.0, the radius equals the focal length, and for germanium (at an index of 4.0), the radius is three times the focal length. For an equiconvex, we neglect the thickness and simply use a value of $2(n-1)(\text{EFL})$ for the radius. This works out to the radius equal to the focal length for an index of 1.5, equal to twice the focal length for an index of 2.0, and six times the focal length for germanium.

An achromatic doublet can easily be drawn in a planoconvex (or planoconcave) shape. The positive element (of the planoconvex doublet) is drawn as an equiconvex lens with radii the same as the radius of the planoconvex singlet cited above, e.g., half of the doublet focal length for glass elements. This is a crude approximation, but it suffices for our purposes. Several examples of these sketches are shown in Fig. 5.4.

The components should be drawn at the diameters necessary to pass the rays of the system. A paraxial raytrace [using Eqs. (1.19) and (1.20)] of the axial marginal ray and of the principal ray will yield y

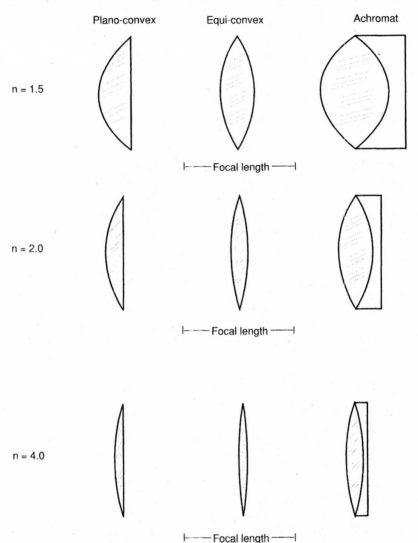

Figure 5.4 Sample quick sketches of singlets and doublets for several values of the index of refraction. All lenses are drawn to the same focal length and diameter for easy comparison. For the planoconvex lenses (both singlets and doublets) the surface radius equals the focal length times $(n-1)$. For the equiconvex form the surface radius is equal to the focal length times $2(n-1)$.

and y_p for each component. Then the diameter necessary to pass the *axial* bundle is simply equal to $2y$. The diameter needed to pass the oblique bundle is equal to $2(y_p \pm V \cdot y)$, where V is the vignetting factor for the oblique bundle. The largest of these diameters is the one to use. The rays (axial and vignetted oblique) should be drawn into the system sketch in order to get a better picture of the way the system functions and also to avoid impossibilities. Section 1.9, and specifically Eqs. (1.23) through (1.26), can be used to generate the oblique bundle ray data from the data of the axial and principal rays.

If the lenses look too fat to you, the odds are that later on the lens designer will have to split them up in order to get a good-quality image. You can get a feel for what the system may look like by splitting them yourself; simply draw the lenses with radii increased by a factor equal to the number of elements you split them into. If you need to split a lens in two in order to make it look reasonable, double the radii of your initial sketch. If you need to split it into three, triple the radii, etc. What you will wind up with is a crude, simple representation of what the system *might* look like. The following paragraphs and the next chapter can help you to decide what sort of a lens design (i.e., singlet, doublet, anastigmat, etc.) will be needed for each component.

For each component, calculate the relative aperture (f-number) in two ways, one by dividing its focal length by the diameter determined two paragraphs above and the other by using the diameter indicated by the axial marginal ray. The latter is an indication of the speed (f-number) at which it must form a decent image; the former indicates the diameter at which it must be constructed.

Similarly, we need to estimate the angular field that the component will have to cover. Again, we can get two values. The first is simply the angle that a ray from the edge of the object (*the object for the component,* not the object for the entire system) through the center of the component makes with the axis; this field angle has to do with the construction, vignetting, etc. Note that this is not the slope of the principal ray, because the principal ray may not go through the center of the component. The second field of view is determined by dividing the smaller of the object or image heights (for the component) by the component focal length.

As an example, consider the erector (component b) of the lens-erecting telescope discussed in Sec. 5.3, and shown in Fig. 5.3. Here the objective lens (component a) forms an image of the field, which is then reimaged by the erector. The (first) half-field of view is the height of the image formed by the objective, divided by the object distance s for the relay. The other (second) half-field is the same image height divided by the component focal length f_b. When the object is at infinity, both field values are the same, but for a system with finite conjugate distances, as is the case for this erector lens, the second field will be larger than the first. This is the field angle to use in deciding what type of lens design form is apt to be able to cover your system's field of view. The reason for this is that the basic field curvature of a lens, called the Petzval curvature, is determined by the image *size,* rather than the field *angle,* and the second field angle described above gives a measure of this relationship which is closer to what most people describe as the "coverage" angle of a lens system.

These field angles and f-numbers can be used to get a preliminary idea of what sort of optics will be needed to make your system work, perhaps by referring to Fig. 6.13.

5.6 Chromatic Aberration and Achromatism

While the intent of this book is to discuss the first-order, or paraxial, aspects of optical system layout, and not to get into the lens design aspects of aberration correction, the chromatic aberrations are really first-order properties in that they can be described in first-order terms. Thus we briefly treat the subject at this point.

The chromatic aberration contribution of a single thin element to the final image of a system can be expressed as follows:

Transverse axial chromatic contribution:

$$\text{TAchA} = \frac{y^2\phi}{Vu_k'} \tag{5.8}$$

Lateral color contribution:

$$\text{TchA} = \frac{yy_p\phi}{Vu_k'} \tag{5.9}$$

where y is the axial marginal ray height, y_p is the principal ray height, ϕ is the element power, V is the Abbe v-number of the material, and u_k' is the slope of the axial marginal ray in the image space. The sum of the contributions from all the elements is the amount of the aberration in the final image. The longitudinal axial chromatic is TAchA/u_k'; this is the distance from the long-wavelength focus to the short-wavelength focus. The lateral color (TchA) is the difference between the image height for short-wavelength and that for long-wavelength light. Lateral color is sometimes expressed as chromatic difference of magnification (CDM), which is TchA/h, where h is the image height. The Abbe v-number, or reciprocal relative dispersion, is $(n_d-1)/(n_F-n_C)$ for visible light; for other spectral regions the v-value uses the index for middle, short, and long wavelengths instead of the index for the d, F, and C lines. Note that mirror elements do not have chromatic aberration.

The entire optical system can be achromatized by arranging it so that the sum of all the element contributions, as given by Eqs. (5.8) and (5.9), add up to zero.

A component can be achromatized as a thin achromatic doublet with the element powers of

$$\phi_a = \frac{\Phi V_a}{V_a - V_b} \tag{5.10}$$

$$\phi_b = \frac{\Phi V_b}{V_b - V_a} \tag{5.11}$$

where ϕ_a and ϕ_b are the element powers and Φ is the power of the doublet. If element a has a v-value of 60 and element b has a v-value of 30 (these are fairly typical values for a glass doublet), it's apparent that the power of element a will be twice that of the doublet, and that element b will have a negative power equal to that of the doublet. The total amount of element power in an achromatic doublet is thus about triple the power of the doublet.

5.7   Athermalization

For most optical systems the chief effect of a temperature change is apt to be a change in the location of the image; in other words, the system goes out of focus. There are three major factors which affect this problem:

1. The thermal expansion of the lens material causes the radii and thickness of each element to increase as the temperature rises; this causes the focal length of an element to increase with temperature. The expansion coefficient for optical plastics is several times that of optical glass.

2. The index of refraction of the lens material changes with temperature. For most (but not all) optical glasses the index rises with temperature, causing the focal length to become shorter. Optical plastics, however, not only have index coefficients which are much larger than that of glasses, but the coefficients are negative, thus guaranteeing an increase in focal length with rising temperature.

3. The structure of the lens mount expands with temperature, changing not only the size of the airspaces between elements but also the lens-to-sensor (film, CCD, or whatever) distance.

The change in power of a thin element with temperature is given by

$$\frac{\delta\phi}{\delta\phi} = \phi\left[\frac{\delta n/\delta t}{(n-1)} - \alpha\right] \tag{5.12}$$

where $\delta n/\delta t$ is the differential of index with temperature and α is the thermal expansion coefficient for the lens material. For a thin doublet we can write

$$\frac{\delta\Phi}{\delta t} = \phi_a T_a + \phi_b T_b \tag{5.13}$$

where Φ is the doublet power, and

$$T = \frac{\delta n/\delta t}{(n-1)} - \alpha \tag{5.14}$$

Athermalization consists of balancing all of these factors so that the temperature effects cancel each other out. For example, we can solve Eq. (5.13) for the element powers to make a doublet with whatever thermal focus shift that we require

$$\phi_a = \frac{\delta\Phi/\delta t - \Phi T_b}{T_a - T_b} \tag{5.15}$$

$$\phi_b = \Phi - \phi_a \tag{5.16}$$

If the thermal focus shift equals the thermal change in the lens-to-sensor distance, then the system is athermalized.

5.8   Sample System Layout

As an example of the layout process, we use an infrared system which requires an external cold stop and also passive thermal compensation. The specifications are listed below:

"First-order" specifications:

1. Focal length: 150 mm

2. Overall length: 260 mm

3. Back focal length: ≥23 mm (minimum clearance)

4. Cold stop location: 17 mm from detector array

5. Packaging: one mirror fold required

"Other" specifications:

6. Thermal compensation: passive

7. Aperture diameter: 31 mm

8. Vignetting: none

9. Field of view: 3.0° ($\pm 1.5°$)

10. Wavelength: 4.8 $\pm$ 0.3 μm

11. Image quality: 50 percent within 50 μm in central 1.5;°50 percent within 75 μm in outer field

12. Distortion: $\leq$2 percent

The requirement for an external cold stop (4) means that the exit pupil of the system must coincide with the cold stop, which is a refrigerated aperture that prevents the infrared-sensitive detector from "seeing" any of the warm, infrared-emitting "structure" of the system. In order to get an external pupil, the system must have at least two separated components, with the second acting as a relay lens. The focal length of a system of this type will be negative (as discussed in Sec. 2.4, where the combined focal length of the objective and erector of the terrestrial telescope was noted to have a negative focal length). Figure 2.7 also shows a pupil at the "glare stop" external to the objective-erector combination.

Five first-order requirements are listed above, and, in a two-component system, we have only two powers and a spacing as variable parameters with which to satisfy these specifications. We approach this optimistically, electing to control the three characteristics which have definite specifications, namely, the focal length, the length, and the cold stop position. Our optimism resides in the hope that when we have these three in hand, the other two (which are the minimal clearance distance and enough space somewhere for a mirror fold) will fall into place. Should this turn out to be overoptimistic, we have to resort to compounding the components, that is, making a component out of two separated subcomponents in the form of a telephoto or retrofocus (Sec. 2.9). Obviously, this is the same as using more than two components.

Going back to Chap. 1, Sec. 1.10, which dealt with systems of two separated components, Eqs. (1.27) and (1.29) give us expressions for the focal length and back focus of a two-component system. Thus, with reference to Fig. 5.5, we have for the focal length

$$\frac{1}{\text{EFL}} = \phi_a + \phi_b - D\phi_a\phi_b \qquad (5.17)$$

and for the back focus

$$B = (1 - D\phi_a)\text{EFL} = L - D \qquad (5.18)$$

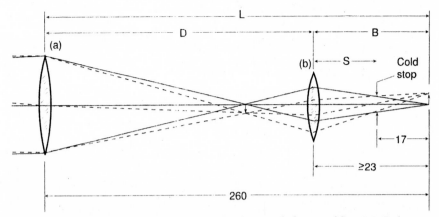

Figure 5.5 Schematic of the optical system for the sample layout of Sec. 5.8. In longer-wavelength infrared systems the detector must not "see" the structure of the device, because even at ordinary temperatures a significant amount of infrared radiation is emitted and can affect the performance. The cold stop is a refrigerated aperture stop which shields the detector.

Our cold stop must be located at the image of the objective aperture which is formed by component b. The object distance from b is $(-D)$ and Eq. (1.4) can be solved for the pupil distance to get

$$S = B - 17 = \frac{D}{D\phi_b - 1} \qquad (5.19)$$

Thus we have three equations in three unknowns, and a simultaneous solution would give us the values of ϕ_a, ϕ_b, and D necessary to satisfy the requirements for the focal length EFL $= -150$ mm, the system length $(L = D + B) = 260$ mm, and the cold stop (pupil) position $(B - S) = 17$.

But we already have a simultaneous solution of Eqs. (5.17) and (5.18) in the form of Eqs. (1.31) and (1.32), which are easily converted to

$$\phi_a = \frac{F - B}{DF} \qquad (5.20)$$

$$\phi_b = \frac{L - F}{DB} \qquad (5.21)$$

where $L = B + D$ and $F = $ EFL. If we determine ϕ_b from Eq. (5.21), we can then determine the stop position from Eq. (5.19).

Using $L = 260$ and $F = -150$ per the specification list, and realizing that $B = L - D = 260 - D$, we choose D as our free variable. We select a few reasonable values for D, evaluate Eqs. (5.21) and (5.19), and tabulate the results as follows.

$D = 230$	$B = 260-D = 30$	$\phi_b = 0.0594203$	$S = 18.16$	$B-S = 11.84$
$= 220$	$= 40$	$= 0.0465909$	$= 23.78$	$= 16.22$
$= 210$	$= 50$	$= 0.0390476$	$= 29.17$	$= 20.83$

Interpolating between $D = 220$ and $D = 210$, we get

$D = 218.3$	$B = 41.7$	$\phi_b = 0.0450396$	$S = 24.72$	$B-S = 16.98$

which gives us a value of 16.98 mm for the pupil/cold stop to detector distance, in good agreement with the 17 mm required by specification 4, and finally we get $\phi_a = +0.0058543$ from Eq. (5.20).

Tracing the axial and principal rays through the system gives us the component diameters required for the specified zero vignetting as $2(|y|+|y_p|)$. For component a, specification 7 sets the clear aperture at 31 mm, and the raytrace yields 20 mm as the necessary clear aperture for component b. These seem reasonable for the component powers we have arrived at.

Silicon is a reasonable material for a system in the specified wavelength region. It has an index of $n = 3.427$ and a v-value of 1511 over our spectral bandpass (specification 10). Using Eq. (5.8) we can calculate the image blur due to chromatic aberration as

$$\Sigma TAchA = \frac{\Sigma y^2 \phi}{Vu'_k} = 0.014 \text{ mm}$$

Given the 50-μm-diameter blur specified for 50 percent of the energy in the image (in specification 11), we arrive at a preliminary conclusion that we may not need to achromatize the system.

However, we are not so lucky with regard to the thermal change in focus. Using silicon, with $n = 3.427$, $\alpha = 2.62e-06$, and $\delta n/\delta t = 159e-06$, we calculate T for Eq. (5.12) as $6.289e-05$. Thus the power of an element when the temperature is raised by 100°C is given by

$$\phi_{100} = \phi(1 + 100T) = 1.00629\phi$$

If the mounting structure of the system is aluminum with a CTE of $\alpha = 0.000024$, we have as the parameters for the nominal system, and for the system at $\delta t = +100$°C:

	Nominal	At $\delta t = +100°C$
ϕ_a	+0.0058543	×1.00629 = +0.0058911
D	218.3	×1.00240 = 218.82392
ϕ_b	+0.0450396	×1.00629 = +0.0453229
B (the space)	41.7	×1.00240 = 41.8006

Using the 100° powers and spacing, and calculating the back focus from Eqs. (5.17) and (5.18), we get $B_{100} = 40.0858$, indicating a thermal focus shift of $(40.0858 - 41.8006) = -1.715$ mm away from the detector, which is at a distance of 41.8006 mm. Our system, with an aperture of 31 mm and a focal length of 150 mm, has an f-number of 150/31 or $f/4.8$, so the blur resulting from the thermal defocus is $1.715/4.8 = 0.35$ mm, many times larger than the 50-μm size indicated in specification 11.

As indicated in Eq. (5.13), we can athermalize a component by combining two materials with different T numbers, much as we achromatize by combining materials with different v-values. Ideally, we would like to find a pair of materials where such a doublet would be both achromatized and athermalized. Some materials suitable for our spectral region are tabulated below.

Silicon	$T =$ 6.29e − 05	$V = 1511$	$1/V =$ 6.62e − 04
Germanium	13.22e − 05	673	14.86e − 04
Amtir	2.93e − 05	642	15.04e − 04
Zinc selenide	3.03e − 05	342	28.42e − 04
Zinc sulfide	3.57e − 05	915	11.15e − 04

When T is plotted vs. $1/V$, as in Fig. 5.6, a line drawn between the points for two materials and extended to the T-axis will indicate the equivalent T-value of an achromatic doublet made from the two materials. The dashed line in the figure indicates that silicon and germanium make an interesting pair. Since the chromatic aberration of the silicon-only system calculated above seemed acceptable, we can consider the possibility of athermalizing the entire system by adding a single negative germanium element, rather than achromatizing and athermalizing both of the components separately. Component a is the bigger contributor to our problems, so we should probably add the germanium element there.

An algebraic solution is possible, but proceeding numerically, we elect to add a germanium element of power ϕ_c to the first component and adjust ϕ_a to maintain the total power of the first component at

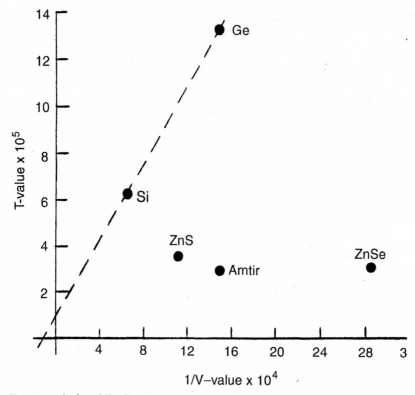

Figure 5.6 A plot of $T = [(\delta n/\delta t)/(n-1)-\alpha]$ vs. $1/V$ for the materials of a system can be used to assess the combined requirement for achromatism and athermalism. A line drawn between the points representing two materials and extended to the T axis indicates the thermal power change $\delta\Phi/\delta t = T\Phi$ for an achromat made of the two materials. If extended to the $1/V$ axis, the intersection can indicate the chromatic aberration of an athermal doublet.

the original value of $+0.0058453$. We make a number of trials using values for ϕ_c of -0.005, -0.010, etc., and find that a value of $\phi_c = -0.011599$ (combined with $\phi_a = +0.017453$, to maintain the first component power at $+0.0058453$) gives us a system with zero thermal focus shift, and a chromatic aberration blur of only 0.008 mm, which is an improvement over the uncorrected value of 0.014 mm which we previously calculated.

The final (lens-designed) configuration is shown in Fig. 5.7. Component b, the relay, was split into three elements by the lens designer. This was necessary to achieve the specified image quality

Figure 5.7 The final lens-designed optics of the sample layout exercise of Sec. 5.8. The objective is a doublet of silicon and germanium which athermalizes the system and partially corrects the chromatic aberration. The relay component was split into three elements of silicon in order to control the distortion and the aberrations of the pupil. The plano elements are the Dewar window and a filter.

and (especially) to correct the distortion and the pupil aberration. The 50 percent blur spot was less than 27 μm over the entire field, and the distortion was less than 1.5 percent, easily meeting the specifications and leaving ample room for fabrication tolerances.

6

Getting the Most
Out of "Stock" Lenses

6.1 Introduction

This chapter is intended as a guide and an aid to the use of "catalog" or "stock" lenses, that is, lenses that can be bought from one of the many suppliers to the trade (or perhaps lenses of unknown origin which you may find in the back of a desk drawer or in a dusty cabinet in your lab). The ideas behind this topic are that (1) there is often a "best" way to utilize a lens for a particular application, (2) various types of lenses have different capabilities as to their coverage and speed, and (3) there are a number of simple ways to measure and test lenses which do not necessarily require a lot of costly equipment.

6.2 Stock Lenses

The benefits of utilizing stock lenses (as opposed to having optics custom-made) are obvious to most of us. Cost is probably the first advantage that comes to mind. Although stock optics are priced at retail, with the substantial mark-up that this implies, this is significantly offset by the fact that stock optics are made in larger quantities than when one or two sets are made to order. The second big advantage is time. The optics are, almost by definition, usually in stock and available for immediate delivery.

Of course, there are drawbacks. The most obvious is that the stock lens has not been specifically designed for the application for which you are using it and cannot be expected to represent the ultimate in performance. Despite this, many systems assembled from stock lenses have turned out to be entirely satisfactory for their intended applications. Many stock lens systems have also been at best only crude prototypes and eventually have had to be replaced by custom designed and fabricated optics. This is not to imply that the latter were without value; much can be learned from a proof of concept or a rough prototype. Our aim here is to make our stock lens system as good as the available stock lenses will allow.

Another drawback with catalog optics is the need to select the optics from a limited list of available diameters and focal lengths, although, if one has an extensive set of suppliers' catalogs, this may be less of a problem than it would appear at first glance.

Many vendors now include the nominal prescriptions for their optics in their catalogs, and many optical software providers include these prescriptions in their data base. When this data is available, the performance of the stock lens system can be evaluated (using the software) without having to make a mock-up. Unfortunately, however, many lenses are sold without construction data. In some cases (particularly for more complex lenses such as anastigmats, microscope optics, and camera lenses) construction data are regarded as proprietary, and the prescriptions aren't released even to large-volume OEM customers. In other cases the vendor simply may not know the prescription data; this is usually the case where the optics are salvage, scrap, or overruns.

An often overlooked problem with "stock" optics is the very real possibility of a limited supply. This is obviously to be expected in the case of salvaged or overrun lenses, where the total supply is often limited to stock on hand. Since vendors occasionally share the same inventory on a cooperative basis, it is wise to be sure that you're not counting the same lot of lenses twice when you check on the available quantity of the optics you plan to use. Another possibility is that the vendor may decide to drop your lens from the line or even retire from the business. If you go into production on an instrument which incorporates stock lenses, it's nightmare time when you discover that there aren't any more.

An inexpensive insurance policy against this problem is to squirrel away a few sets of optics, so that when the ax falls you still have a reasonable chance of survival. A skilled optical engineer with good laboratory equipment can measure the radii, thickness, spacings, and index of a sample lens. Even if the measurements are not too accurate, an optical design program can optimize or touch up the measured data to suit your purposes. Then you can have some more fabricated. Even if the vendor publishes the construction data of a lens, it's still wise to hold on to a couple of samples; the published data may or may not be accurate.

In the case of salvage optics, you have probably ordered from a catalog listing of diameter and focal length. You should be aware of a couple of factors. The listings are probably based on measurements made on a sample lens, and the listing probably gives the focal length and diameter to the nearest millimeter. There may be more than one lens in the vendor's stock which fits the diameter and focal length description to this accuracy. The next time you order the same catalog item you may get a different lens, and this lens may perform differently.

6.3   Some Simple Measurements

This section is written for the benefit of the reader who does not have access to the usual optical measurement and laboratory equipment. A lens bench with collimator and a measuring microscope are essential for really accurate measurements of the imaging (i.e., gaussian) properties of a lens. However, there are a number of simple ways that an approximate measurement of focal length and back focal length can be made.

The measurement of the *back focal length,* or BFL (see Sec. 1.2 and Fig. 1.1), is relatively easy. When using a lens bench, a target at infinity is provided by the collimator; the location of the focal point is determined by focusing the lens bench microscope alternately on the focal point and on the last surface of the lens. A distant object (a tree, a building, a telephone pole) makes a reasonable substitute for a collimator. To get a rough measurement, one simply focuses the target on a light-colored wall or a piece of typing paper as shown in Fig. 6.1,

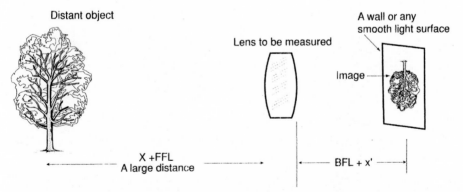

Figure 6.1 The back focal length (BFL) of a lens can be measured using a distant object (instead of a collimator) and measuring the lens-to-image distance with a scale or ruler. If the object is not effectively at infinity, the newtonian focus shift ($x'' = -f^2/x$) is subtracted from the measurement.

and measures the lens-to-image distance with a scale. If the lens has spherical aberration (and most simple lenses do) your measurement will come up a bit shorter than the paraxial back focus. If there is enough light, the spherical aberration can be minimized by reducing the lens aperture with a mask.

Using an object which is not collimated (i.e., not located at infinity) will introduce a small error in your measurement. The error in locating the focal point is indicated by Newton's equation [Eq. (1.1)] as $x' = -f^2/x$, where x' is the error, f is the focal length, and x is the distance to your target. For example, if you measure the back focus of a 2-in-focal-length lens using a target down the hall which is only 50 ft (600 in) away, the error will be $x' = 2^2/600 = 0.007$ in; this is less than the error in your crude measurement of the back focus is likely to be. Even in cases where this error is significant, you can always calculate the error and subtract it from the measured BFL to improve the validity of the result.

One problem associated with this technique is that stray light falling on the image will make it difficult to see and focus. A cardboard screen with a hole for the lens is one way around this problem; using a light bulb as a target in a darkened room is another. Figure 6.2 shows a matchbox slipped over the scale and slid into best focus as a handy tool for making the measurement.

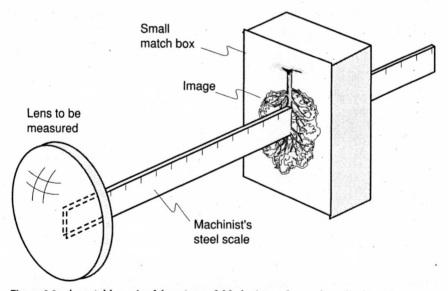

Figure 6.2 A matchbox-sized box (or a folded piece of pasteboard) slipped over a machinist's steel scale makes a convenient way to measure the back focal length of a lens.

The measurement of the *effective focal length* (EFL) is considerably more difficult, because it involves the location of the principal points. Figure 1.2 shows the principal point locations for simple lenses of various shapes. A fair estimate can be made for single elements and most cemented doublets by assuming that the space between the principal points is approximately one-third of the axial thickness of the lens. [A somewhat better estimate for singlets is $t(n-1)/n$.] For planoconvex forms, one principal point is always located at the curved surface; for equiconvex lenses the points are evenly spaced within the lens. As illustrated in Fig. 6.3, adding a suitable fraction of the axial thickness to the measured BFL will then get an estimate for the EFL. For an anastigmat, such as a Cooke triplet or a Tessar, the principal point locations are more difficult to estimate. Adding one-half to two-thirds of the vertex length of the lens to the BFL is about the best estimate one can make. In a complex lens the principal points are often almost coincident, and occasionally reversed.

To get a better value for the focal length one must measure a magnification. Magnification is simply the ratio of the image size to the object size. An illuminated object of known size is imaged and the image size is measured; admittedly, doing this accurately may be easier said than done, but it can be done. The setup is shown schemati-

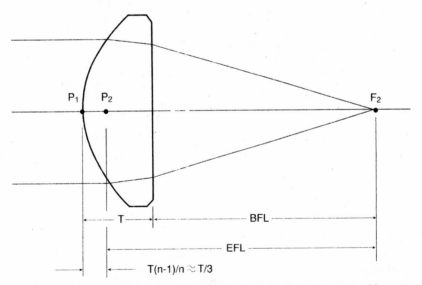

Figure 6.3 One way of estimating the effective focal length (EFL) is to add a suitable fraction of the element thickness to the measured back focal length (BFL). For a planoconvex element, the convex side should face the distant target to minimize spherical aberration. If the lens is reversed, the measured BFL will equal the EFL, but the spherical aberration will be much larger and will affect the measurement.

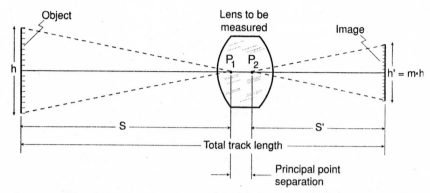

Figure 6.4 Setup for measuring effective focal length (EFL) by measuring the magnification at finite conjugates: The magnification (here taken as the ratio of measured image size to object size—positive for an inverted image) and the track length are measured. Then the gaussian conjugates can be found by dividing the total track length (minus the principal point separation) by the magnification plus 1 to get S. Then $S'' = mS$, and the Gauss equation can be solved for the focal length.

cally in Fig. 6.4. The object-to-image distance (often called the total track length) is measured. The estimated spacing between the principal points is subtracted from the track length. This adjusted length is scaled to get s and s'. Dividing the adjusted track length T by $(m+1)$, where m is the ratio of the image to object size will get s; then $(T-s)$ equals s'. (Note that we use a positive sign for the magnification m in this case.) Now we solve the Gauss equation [Eq. (1.4)] to get the focal length.

If this process is carried out accurately for several different object-to-image distances, it is possible to eliminate the estimation of the principal point separation by making a simultaneous solution for the exact value.

Sample calculations

The object is a back-illuminated transparent scale, 15 in long, and the lens forms an image which is measured at 3.5 in. The magnification is thus $m = 3.5/15 = 0.2333$. The object-to-image distance is 40 in, and the lens is 1 in thick. Assuming that the principal points are separated by one-third the lens thickness, or 0.333, the adjusted track length is 39.667 in. We get the Gauss object distance by dividing the track length by $(m+1)$, or 1.2333, which gives us $s = 39.667/1.2333 = 32.162$ in and $s' = 39.667-32.162 = 7.5045$ in. Substituting s and s' into Eq. (1.4), and solving for f gives us the focal length. (Note that our sign convention requires s to be negative.)

$$\frac{1}{s'} = \frac{1}{s} + \frac{1}{f}$$

$$\frac{1}{7.504} = \frac{1}{-32.162} + \frac{1}{f}$$

$$f = 6.0847 \text{ in}$$

Yet another way to measure EFL involves the use of a distant target whose *angular* size is known. The size of the image of the target is measured. Then the focal length of the lens is equal to half the image size divided by the tangent of the half angle which the object subtends. If the object is not at infinity, the Newton correction of f^2/x can be applied (as in the discussion of BFL above). A distant building with vents, chimneys, elevator towers, etc., on its roof line can be measured with a theodolite through a convenient window to serve as a target of this type. Alternatively, the measurement of a lens of known focal length can be used to determine the angle.

6.4 System Mock-up and Test

A simple way of testing an optical system is to make a mock-up by fastening the elements to a ruler (or other convenient straightedge) with one of the available waxes which are commonly used for this purpose. This material is basically beeswax, formulated so that, when warmed in the hand, it is soft, pliable, and adherent to most things. But it becomes relatively firm at room temperature. This wax is available in stick form as optician's "red wax" or in an unpigmented tan color. Universal Photonics, Inc., of Hicksville, NY, carries "Red (or White) Sticky Wax" in stick or bulk form, and Central Scientific has a tacky wax packaged in 1- or 2-lb cans. The optics are conveniently stuck on the edge of the scale as indicated in Fig. 6.5, and a fair impression of the performance can be obtained by viewing through the optics or projecting the image on a suitable screen.

A somewhat more elaborate lens bench can be made inexpensively from cold-rolled hexagonal bar as shown in Fig. 6.6. Two bars are fastened parallel to each other, spaced apart so that short carrier sections can slide along the length of the bars. Each carrier section is drilled to accept a vertical rod, which is adjustable for height and fixed by tightening a set-screw. The lens can be waxed on the end of the rod, or a short length of angle iron can be threaded on the end of

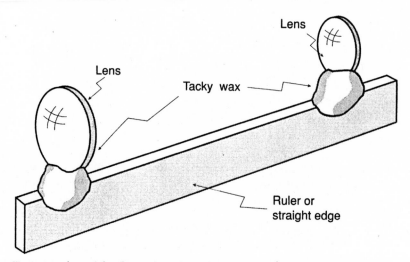

Figure 6.5 A straightedge and some tacky wax make a convenient way to mock up an optical system. This is especially handy for trying out visual systems such as telescopes or microscopes.

the rod to serve as a V block to hold the optics. This bench allows easy and rapid adjustment of the spacings and alignments of the system components.

In making a mock-up of a system, the alignment of the optics with respect to the optical axis can be extremely important. Establish an axial center point on the object, and add the downstream components one at a time, making sure that each image they form is well centered on the axis. Visually sighting over (or through) the optics is often helpful, as is the use of an HeNe laser beam as an alignment aid. Be especially careful in the adjustment of mirrors and prisms; many people make the mistake of underestimating how serious the effect of a misaligned reflecting component can be. If cylindrical elements are included in the system, the orientation of the cylinder axes is critical, especially if there are orthogonal cylinders or if the object is a slit.

The simplest method of performance testing is by the visual evaluation of resolution. As indicated in Sec. 4.3, resolution is not the be-all, end-all of image evaluation, but it is easy and quick to test. A bar target similar to that shown in Fig. 6.7 is readily obtainable; one can be "homemade" with black drafting tape and white paper, or a copy of USAF 1951 can be purchased for a few dollars (or Fig. 6.7 can be xeroxed and used). Another type of test target can be made by scratching fine lines through the aluminized surface of a first-surface mirror, and illuminating it from the rear. The image of a pinhole tar-

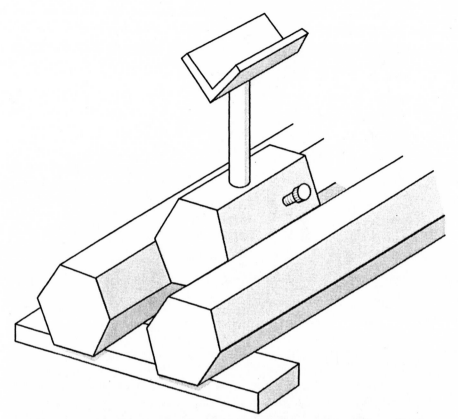

Figure 6.6 An inexpensive lens bench can be made from hexagonal steel bars. Two bars are fastened together as shown here so the short sections of bar will slide between them. The optics can be waxed on the sliders, or angle iron V troughs threaded on rods which are adjustable for height alignment can be used to support the optics.

get can be examined to analyze both aberrations and alignment problems (which are indicated by a nonsymmetrical blur spot on the axis).

For testing eyepieces, magnifiers, or similar devices, where the appearance of an extended field is important, a piece of graph paper makes an excellent test target. One can readily evaluate the image distortion and curvature of field, as well as the effect of changing the eye position, with this sort of target.

An often overlooked factor in developing an optical system is the deleterious effect of stray light. This is light from outside the field of view which is reflected or scattered from some part of the assembly (typically the housing of the optics) into the field of view, where it can either reduce the image contrast or produce ghost images. This can come as a great surprise when the first complete model of the system

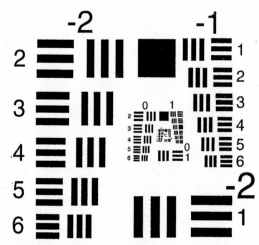

Figure 6.7 A three-bar resolution target. Each pattern differs from the next by a factor equal to the sixth root of 2 (1.1225), and thus the groups of six differ by a factor of 2. Targets of this type are commercially available on film or as metal evaporated on glass.

is made, because often the system mock-up is mechanically quite different from the final product, and the stray light may not be present in the mock-up. There are two ways that stray light can be handled. In a system which has an internal pupil, a glare stop, as shown in Fig. 2.7, is both invaluable and effective. Stops can be placed at every internal pupil and also at every internal image plane. The other technique is simply blackening the offending (reflecting) member. Sometimes it is difficult to know where the problem is. An effective method of locating the source is to simply look into the optics from the location in the image where the stray light is showing up. In other words, put your eye there and look back into the optics. This view is most sensitive if the eye is placed in a location where the image should be dark, e.g., outside the field of view. Then you can see the (image of) structure from which the light is reflected. Another place to look is at the exit pupil, which can be examined with a magnifying glass. The image of the inside of the optical instrument is typically focused near the exit pupil. Once you've located the culprit, a flat black paint (such as Floquil flat black model locomotive paint, available from your local hobby shop) will usually fix things. Another very effective material is black flocked paper, which can be glued on the reflecting surface. Flocked paper is a good absorber; it can be obtained from Edmund Scientific in Barrington, NJ.

6.5   Aberrations

This section presents a brief straightforward discussion of aberrations (which are imperfections in the image). For a more complete treatment, the texts listed in the bibliography are recommended. The so-called primary aberrations are

1. Spherical aberration
2. Coma
3. Astigmatism
4. Petzval field curvature
5. Distortion
6. Axial chromatic aberration
7. Lateral color

The first five are called Seidel aberrations or third-order aberrations. The image blurs caused by these aberrations vary with aperture y and field h according to

1. y^3
2. y^2h
3. yh^2
4. yh^2
5. h^3
6. y
7. h

These relationships can be used to estimate how the aberration blurs will change if the aperture or field of view of a simple system is changed. For example, aberration 2, coma, has a blur spot whose size varies with y^2h, so that if we double the aperture size the blur will increase by a factor of 4 (2^2), but doubling the field h will simply double the blur. Note that the exponents add up to three for the Seidel aberrations, hence the name "third-order." Chromatic aberrations 6 and 7 vary as y and h to the first power; as you may have deduced from this, these are "first-order" aberrations.

Spherical aberration causes a circular image blur which is the same over the entire field of view. It is the only monochromatic aberration which occurs on the optical axis. As shown in Fig. 6.8A, it results from the rays through different zones of the aperture being focused at different distances from the lens.

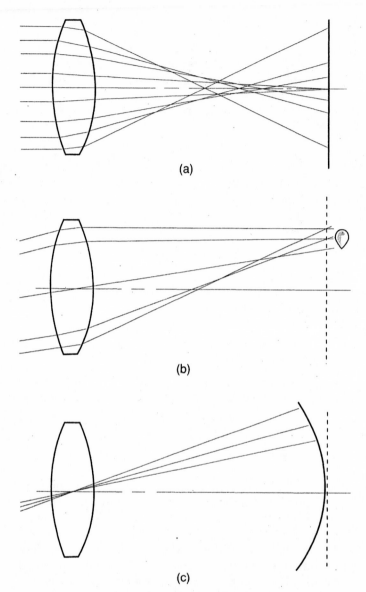

(a)

(b)

(c)

Figure 6.8 The primary aberrations: (A) Spherical aberration—the rays through the outer zones of the lens focus (cross the axis) closer to the lens than the rays through the central zones. This is "undercorrected" or negative spherical aberration. (B) Coma—the rays through the outer zones of the lens form a larger image than the rays through the center. "Overcorrected" or positive coma. (C) Field curvature—images farther from the axis focus nearer to the lens than the on-axis images. This is an "inward" or negative field curvature.

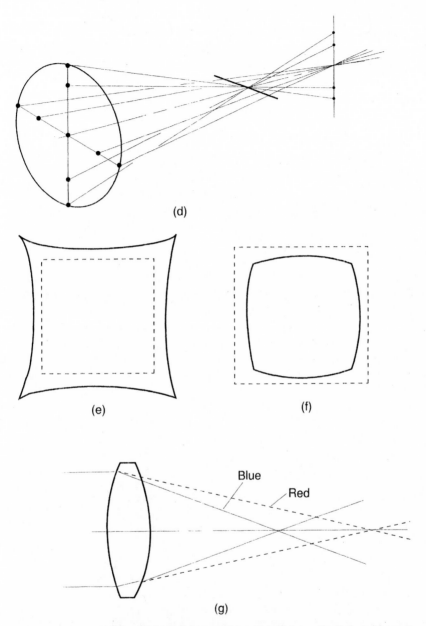

Figure 6.8 (*Continued*) (*D*) Astigmatism—the tangential (vertical) fan of rays is focused to the left of the sagittal (horizontal) fan. This is negative astigmatism. (*E*) Distortion—the magnification increases as the field angle increases. "Pincushion" or positive distortion. (*F*) Distortion—"barrel" or negative distortion. (*G*) Axial chromatic aberration—the short-wavelength (blue) light is focused nearer the lens than the long-wavelength (red) light. "Undercorrected," negative chromatic.

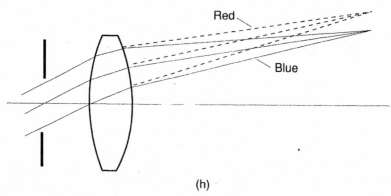

(h)

Figure 6.8 (*Continued*) (*H*) Lateral color—the image size is larger for long-wavelength (red) light than for short-wavelength (blue) light. Negative lateral color.

Coma causes a comet-shaped flare for off-axis points. It results from a lens having different magnifications for rays passing through different zones of the aperture. Coma is illustrated in Fig. 6.8*B*. The size of the flare increases with the distance of the image from the axis. The coma flare always points toward or away from the axis.

Astigmatism and field curvature cause the off-axis images to be focused on a curved, saucer-shaped surface as drawn in Fig. 6.8*C*, instead of an ideal flat image surface. This surface is paraboloidal, so the amount of defocusing, and the blur it causes, vary as the square of the image distance from the axis. The *astigmatism*, illustrated in Fig. 6.8*D*, causes the image of a line pattern which has a radial direction to be focused on a differently shaped saucer than a line pattern running in a perpendicular direction.

Distortion causes straight lines which do not intersect the axis to be imaged as curved lines. This results from the fact that the magnification varies across the field and causes the image of a square or rectangular object to be bowed outward (barrel distortion, Fig. 6.8*F*) or sagged inward (pincushion distortion, Fig. 6.8*E*).

Axial chromatic aberration is a variation of image position with wavelength and is caused by the variation of index with wavelength. The index of refraction for a short wavelength is higher than for long wavelength, so blue light is focused closer to the lens than red light, as indicated in Fig. 6.8*G*. This causes a blurred image which is the same over the whole field. Axial chromatic and spherical are the only primary aberrations that exist on the axis. Obviously there are no chromatic aberrations if the light is monochromatic, as with laser light.

Lateral color, often called chromatic difference of magnification or CDM, results from a different magnification for each color. As drawn in Fig. 6.8*H,* red and blue images of an off-axis object point are separated in a radial direction, and the separation increases with the distance from the axis.

In addition to aperture and field, as cited above, most aberrations vary with the shape of the lens element. For example, if the object is distant, spherical aberration is a minimum for a particular shape. For glass lenses this shape is approximately planoconvex, with the convex surface facing the distant object. The variation of spherical aberration with lens shape, as a function of index, is shown in Fig. 6.9.

For distant objects, biconvex and meniscus (one side convex, the other concave) lenses have more aberration. But for very high index lenses, such as silicon ($n = 3.5$) or germanium ($n = 4.0$), the minimum spherical shape is meniscus with the convex side facing the object as indicated in Fig. 6.9. But if we consider a different object location, the shape for minimum spherical changes. For example, at one-to-one imagery, the equiconvex shape has the least spherical. Thus for any given element in an optical system, one orientation may be much preferred to the other. Figure 6.10 shows the variation of the angular spherical aberration blur with object distance for three different lens shapes.

The position of the aperture stop, relative to the lens shape, can have a big effect on the off-axis aberrations (coma, astigmatism, distortion, and lateral color). Often, reversing a lens will make a significant difference in the system performance at the edge of the field, because it changes its orientation with respect to the stop.

In general, the higher the power of a lens, the more aberration it will introduce to the system. So a good way to reduce the aberrations is to substitute two low-power elements for a single high-power element as shown in Fig. 6.11. When the elements are properly shaped to take advantage of this "split," the spherical aberration can be reduced by a factor of 5 or so. A split is most effective against spherical aberration. Another technique for reducing spherical is to spread the "work" equally, where work is the amount that the lens bends the light ray. A quick look at Eq. (1.19) indicates that "work" is simply $y\phi$, the product of ray height and lens power.

Petzval field curvature is a function of lens power and index; it is not affected by shape or object distance, as are the other aberrations. In essence, it amounts roughly to the sum of all the positive power in the system minus the sum of all the negative power. Looked at this way, it's apparent that the usual field curvature problem that we encounter when using stock optics is due to having only positive ele-

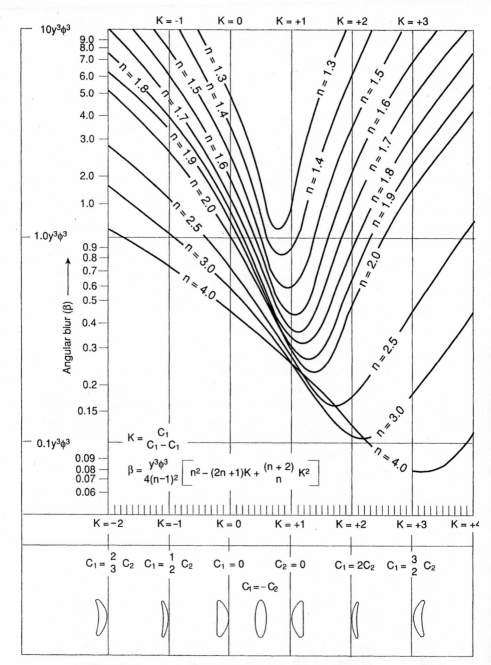

Figure 6.9 The angular spherical aberration blur of a single-lens element as a function of lens shape for various values of the index of refraction. The object is at infinity. ϕ is the element power and y is the semiaperture. The angular blur β can be converted to longitudinal spherical aberration by LA $= -2\beta/y\phi^2$, or to transverse aberration by TA $= -2\beta/\phi$.

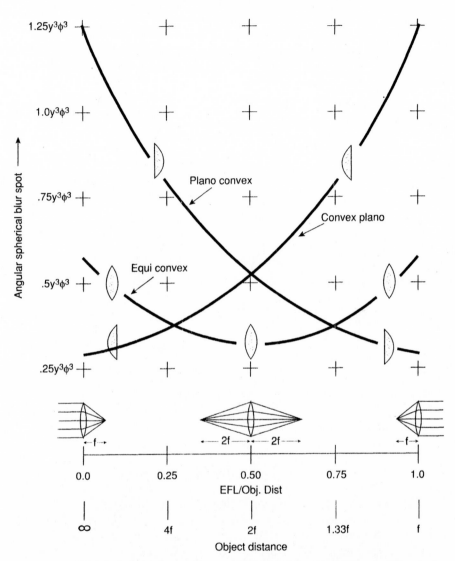

Figure 6.10 Spherical aberration varies as a function of the object distance. The graph plots this variation for three-element shapes for a lens with an index of $n = 1.80$. Note that for the planoconvex shape the minimum spherical occurs when the curved side faces the longer conjugate, whereas for the equiconvex shape the minimum is at 1:1 magnification. The size of the blur spot can be found by multiplying the angular blur by the image distance. ϕ is the element power and y is the semiaperture.

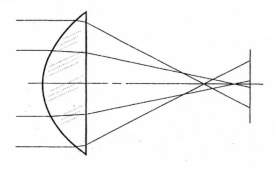

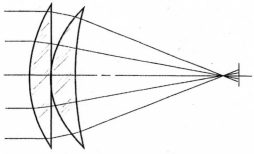

Figure 6.11 The spherical aberration of an element can be reduced by a factor of 5 or more by splitting it into two elements, each shaped to minimize its spherical aberration.

ments in a system; this is why the field curvature is almost always curved inward, toward the lens. Flat-field lenses correct this by introducing negative power where it doesn't have much effect on the ray slope, namely, where the ray height is low. Anastigmat lenses get their correction by using high index glass and separating positive and negative power in order to lower the ray height on the negative elements relative to the positive elements. In a "stock" layout we are usually stuck with lots of positive lenses, but sometimes there is one thing that we can do, and that's the introduction of what is called a *field flattener*. This is usually a negative element placed at or near a focal plane (where the ray height is very low and the lens has little effect on the other aberrations or on the image size) as shown in Fig. 6.12*A*. A field flattener can have a very salubrious effect on a system with too much inward field curvature. Note that the reverse is also true; a positive field lens will increase the inward field curvature.

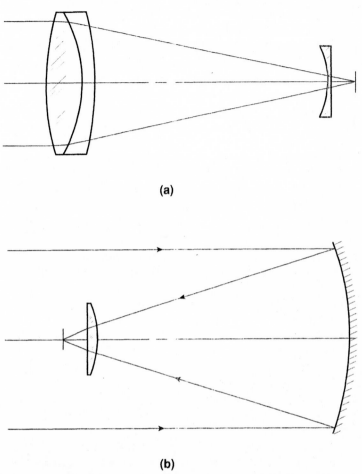

(a)

(b)

Figure 6.12   A field flattener is a lens placed close to the image where it has little effect except on the Petzval field curvature. A negative lens, as shown in (A), will flatten an inward-curving field such as that produced by positive components. In (B), the concave mirror has a backward-curving field which is corrected by a positive field flattener lens.

Note also that while a positive converging lens has inward field curvature, a converging (concave) mirror has backward field curvature and needs a positive field flattener as shown in Fig. 6.12B.

6.6   Capabilities of Various Lens Types

The aperture and field at which a lens is used can be said to define its capabilities. However, it should be apparent that the level of image quality involved will effectively determine how much aperture and field the lens can handle. Thus, a given type of lens will be "capable" of covering a wider field at a larger aperture if the required image quality is low than if it is high. Therefore, we can describe capabilities only in very approximate terms. Your application may be such that a given lens type can do much more or much less than the average.

A single-lens element is generally totally uncorrected and may be afflicted with all of the aberrations. Since the aberrations vary with field and aperture, simple lens systems are best when used at small fields and small apertures. A simple "landscape lens" as used in an inexpensive camera works at $f/10$ or $f/15$ and covers a field of 30 or 40°. At higher speeds the field coverage is usually much less. A doublet lens can be corrected for chromatic aberration (an achromat) as well as spherical aberration and coma, so that it can produce a good image over a small field of view. Typical usage is at about $f/5$ and 1° field. Any *thin* lens system, that is, one with a very short overall length compared to its focal length (such as a single element or a cemented doublet all by itself), will *always* be afflicted with a large amount of astigmatism, which will cause an inward-curving field. Thus a thin system such as a telescope objective can cover only a very small field of view, to the order of a few degrees, if a good image is required. A longer system can, if handled properly, greatly improve the off-axis image quality (even if the length results from simply spacing a stop away from a single element).

Multielement lenses are usually designed for specific applications. Camera lenses (anastigmats) are usually designed for objects at a distance and are designed to cover a relatively wide angular field, often to the order of 40 or 50°. Most such lenses can be depended on to perform well as long as the object is at least 25 focal lengths away. In general high-speed lenses are more sensitive to close object distances than slower lenses. But note that *enlarging lenses* are designed to work at close distances, from ten-to-one down to one-to-one magnification.

Ordinary *microscope objectives* are intended for use at a specific image distance (about 160 mm) and are very sensitive to departures from their nominal conjugates. "Infinity corrected" objectives are designed to be used with the image at infinity, rather than the usual distance. Most microscope objectives are designed to be used with a thin glass coverslip between the lens and the object and will perform poorly without one; there are also objectives designed to be used without a coverslip. The primary effect of a departure from the design conditions is to change the spherical aberration. These comments are especially important for high power and high numerical aperture (NA) microscope objectives, where a small departure from the nominal usage can ruin the performance of the system.

Figure 6.13 shows the typical angular coverage and relative aperture capabilities of many of the common lens types, and can be used as a rough guide to the complexity of lens usually necessary for various applications.

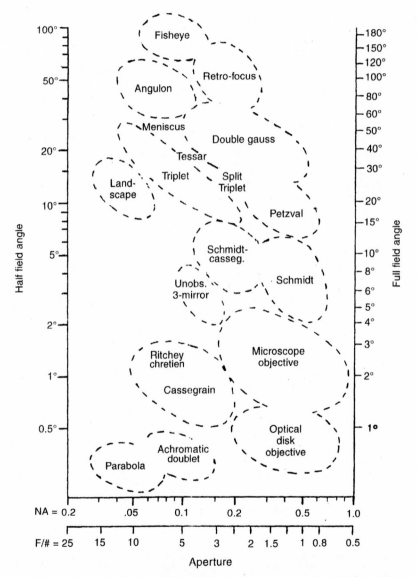

Figure 6.13 Map showing the design types which are commonly used for various combinations of aperture and field of view.

6.7   Unusual Lens Types

An *aspheric surface* can be used by a lens designer as an additional degree of freedom in correcting the aberrations. For example, in a singlet lens an aspheric can be used to correct the spherical aberration. But note well that the presence of an aspheric surface on a lens does not mean that all the aberrations have been corrected. Molded aspheric surfaces are commonly used in projection condenser systems. Diamond turned aspheric lenses for infrared use are available (as stock optics) in materials such as germanium and zinc selenide.

A *fresnel lens,* as shown in Fig. 6.14, is a lens where the thickness has been removed to produce a thin, light piece. Molded glass fresnels have been used for years in spotlights, marine lanterns, railway signal lights, etc., but glass fresnel lenses cannot be molded with fine details, so the step size in glass is quite large. However, plastic fresnels can be made with very fine steps and in very thin configurations. The slope of the step face can be made so that spherical aberration is corrected, and such a lens does a very acceptable job for tasks such as concentrating energy or illumination condensers, but in general fres-

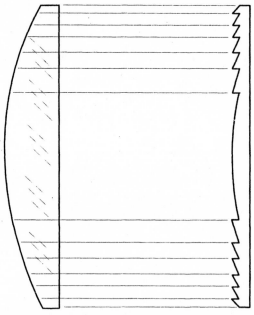

Figure 6.14   The fresnel lens—each facet of the fresnel lens has the slope of the corresponding section of the lens, but the thickness is reduced by its stepped construction.

nels are not used to produce high-quality images. The fresnel is most commonly seen as the condenser of an overhead (transparency) projector, a field lens behind the screen of a projection TV, a field lens in the viewfinder of a single-lens reflex camera, or a wide-angle viewer for the rear window of a van.

A *diffractive* surface (sometimes called a kinoform or binary) is a fresnel surface where the step depth produces a phase (or path length) step of one wavelength, or a small integral number of wavelengths. The fresnel surface can be shaped (or blazed) to correspond to an aspheric surface and can thus correct aberrations. Since the surface is a diffraction device analogous to a grating, its chromatic dispersion characteristics are of the opposite sign from, and quite extreme compared to, ordinary optical material. This allows, for example, a single-lens element to be corrected for chromatic aberration, spherical aberration, and coma. As with most diffraction effects, there are efficiency limitations on bandwidth, etc.

Grin (gradient index) rods or *Selfoc* lenses are soda-straw-sized glass rods with an index of refraction which varies radially, from a high index near the axis of the rod to a low index at the edge. A quadratic variation of the index allows the device to function as a reasonably well-corrected lens. A short section acts like an ordinary lens. Small, short grin rods are often used as objective lenses for the flexible fiber-optic endoscopes used in arthroscopic surgery. As shown in Fig. 6.15, a longer rod can behave like a three-lens system: a relay lens, a field lens, and another relay lens, to produce an upright image at unit magnification. An array of such rods can be found in tabletop copy machines. A still longer grin rod can be used as a periscope or rigid endoscope.

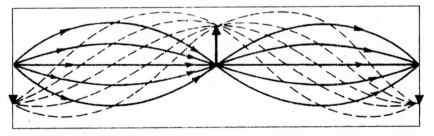

Figure 6.15 A "grin rod" or "Selfoc" lens has a radial gradient index, high at the central axis and lower toward the edge of the rod (according to a quadratic law which causes the rod to act like a lens). In this figure a long rod behaves as if it were three lenses: two relay lenses (which are indicated by the path of the solid-line rays), plus a field lens (as indicated by the dashed rays).

Another type of gradient index is called an *axial index gradient*; here the index changes along the axis as you progress through the lens. When used in a spherical-surfaced lens, the index gradient can cause a spherical surface to refract light like an aspheric surface, and a single-lens element can be made free of spherical aberration and coma.

6.8   How to Use a Singlet (Single Element)

When using "stock" lenses our choice of elements is quite limited. Indeed, the usual choice that optical catalogs offer us is between a planoconvex (or nearly so) lens and a biconvex (which is probably equiconvex) lens, as indicated in Fig. 6.16. In the following discussions, a biconvex which has one surface *much* more strongly curved than the other can be regarded as a planoconvex. If both surfaces are similarly shaped, the lens can be treated as equiconvex (although in this case there may be a preferred orientation).

We can start our considerations by assuming that we have a system which requires a well-corrected image over a small field of view. What this means is that we will want to minimize the spherical aberration in the image and that we won't worry too much about the off-axis image. We consider the three cases diagramed in Fig. 6.17.

For a *telescope objective type,* that is, a system with the object a long distance away, say more than 5 or 10 focal lengths, we choose a planoconvex lens and orient it with the curved surface toward the distant object.

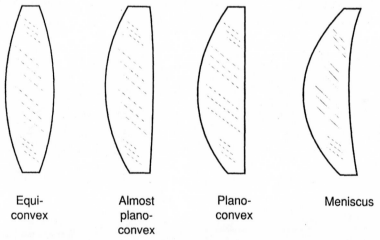

Equi-   Almost   Plano-   Meniscus
convex   plano-   convex
             convex

Figure 6.16 The forms of single-lens elements which are commonly available from catalogs.

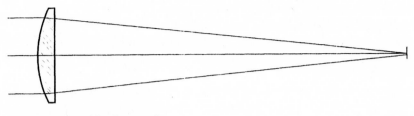

(a) "Telescope objective type"

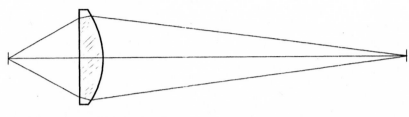

(b) "Microscope objective type"

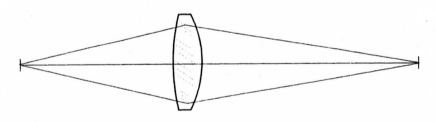

(c) "Relay lens type"

Figure 6.17 For applications with small fields of view there are three common cases: (A) the "telescope objective" type, where the object is a long distance to the left; a planoconvex lens with the convex side facing the distant object minimizes the spherical aberration; (B) the "microscope objective" type, where the image is distant, and the convex side of a planoconvex lens faces the image to minimize spherical; and (C) the "relay lens" type where neither conjugate is greatly longer than the other; a biconvex lens is best, with the more strongly curved surface facing the longer conjugate.

For a *microscope objective type,* that is, a lens which will magnify the object by five or more times, we again choose a planoconvex but face the plano side toward the object.

For a *relay lens type,* that is, a lens with a magnification between $(-)5\times$ and $(-)0.2\times$, a biconvex lens is the choice. If the lens is not equiconvex, orient the more strongly curved surface (i.e., the shorter radius) toward the longer conjugate.

For a wider field of view, we must be more concerned with the off-axis aberrations. In this case the location of the aperture stop can be critical, since its position will affect coma, astigmatism, and field curvature. In general, we must sacrifice the image quality at the axis in order to get better performance at the edge of the field. Usually a planoconvex or a meniscus (one side convex, the other concave) is the best bet, with the aperture stop on the plano (or concave, if meniscus) side of the lens, as shown in Fig. 6.18. Note that the lens wants to sort of "wrap around" the stop. This is the reason that most camera lenses have an external shape which is almost like a sphere with the stop in the center. When there is no separate stop and the object is some distance away, a planoconvex lens with the plano toward the object often works well (because the coma in the image produces a sort of field flattening effect).

When a singlet is used as a *magnifying glass* and held close to the eye as in Fig. 6.19A, a planoconvex lens with the plano side toward the eye works best. Here the pupil of the eye acts as the aperture

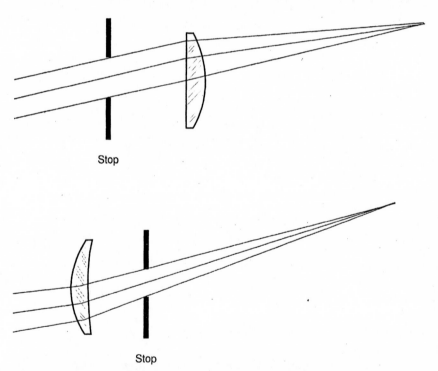

Stop

Stop

Figure 6.18 For applications where a wider field of view is covered, the lens is oriented with the field aberrations (coma and astigmatism) as the prime concern. An aperture stop, spaced away from the lens on the "concave" or plano side (as shown in this figure), can have a favorable effect on the off-axis imagery.

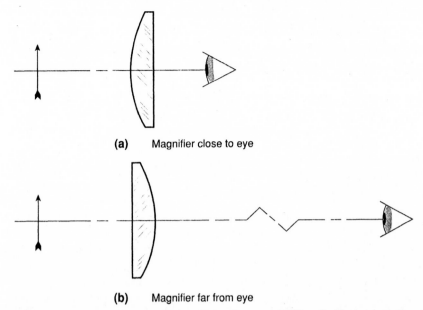

(a)   Magnifier close to eye

(b)   Magnifier far from eye

Figure 6.19 When a planoconvex lens is used as a magnifier, the best orientation depends on the location of the eye, which acts as the aperture stop. When close to the eye, the plano side should face the eye; this orientation minimizes distortion, coma, and astigmatism. When far from the eye, the convex side should face the eye.

stop, and the lens is "wrapped around" it. This usage is the same as found in head-mounted displays (HMD) and is also much like that in a telescope eyepiece. However, if the lens is a foot or two from the eye, as shown in Fig. 6.19B or as in a tabletop slide viewer or in a head-up display (HUD), the plano side should face away from the eye. This is because the image of the eye formed by the lens is a pupil of the system, and, with the lens well away from the eye, this image is on the far side of the lens. We want the plano side to face the stop/pupil, so that the lens wraps around the pupil. For a general-purpose magnifier which is used both near to and far from the eye, an equiconvex shape is probably the best compromise, although the two-lens magnifier as described below is much better.

Note that these comments can also be applied to a planoconvex cemented achromatic doublet.

6.9   How to Use a Cemented Doublet

Most stock cemented doublets are designed to be corrected for chromatic and spherical aberration, and probably coma as well, when used

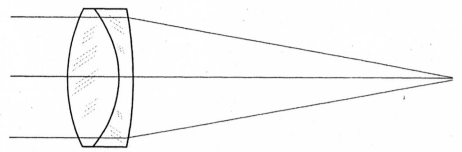

Figure 6.20 Most "stock" achromatic doublets are designed as telescope objectives and are corrected for chromatic and spherical aberration as well as coma with the object at infinity. The more strongly curved surface should face the more distant conjugate.

with an object at infinity. In other words, they are effectively telescope objectives and are designed to cover a small field of view. As illustrated in Fig. 6.20, the external form is usually biconvex, with one surface much more strongly curved than the other, i.e., close to a planoconvex shape. Just as with the planoconvex singlet, the more strongly curved surface should face the distant object. If the doublet is used at finite conjugates, the strong side should face the longer conjugate.

If neither of the exterior surfaces is more significantly curved than the other, the odds are that the lens was not designed for use with an infinite conjugate. It may be corrected for use at finite conjugates, or, what is more likely, it may have been part of a more complex assembly. Here, some experimentation is in order. Try both orientations and observe the performance. Again, as with the singlet, it's highly probable that the stronger surface will want to face the longer conjugate.

A meniscus-shaped doublet is rarely found as a "stock" lens; such a doublet is most likely either surplus or salvage, and its shape results from the design of which it was originally part. Although a (thick) meniscus is very useful as a lens design tool (to flatten the field), such a lens will probably not be too useful in your system mock-up; it might work out as part of an eyepiece, with the concave side adjacent to either the eye or the field stop.

6.10   Combinations of Stock Lenses

Often the use of two lenses instead of one can make a big improvement in system performance. The following paragraphs discuss a number of possibilities.

High-speed (or large NA) applications.   The usual problem in fast systems is spherical aberration. Using two lenses instead of one, with each shaped to minimize the spherical aberration, can alleviate the situation. The optimum division of power is equal; both elements have the same focal length, and the sum of their powers equals the power of the single element which they are replacing. For a distant object, the first element should be planoconvex, with the convex side facing the object. Ideally, the second element should be meniscus, with the convex side facing the first element as sketched in Fig. 6.21A. But since meniscus stock elements are hard to come by, the usual stock lens arrangement is another planoconvex with its convex side also facing the object, as in Fig. 6.21B. If one planoconvex singlet is stronger than the other, place it in the convergent beam. If one of the lenses is a doublet, it should probably be the one facing the dis-

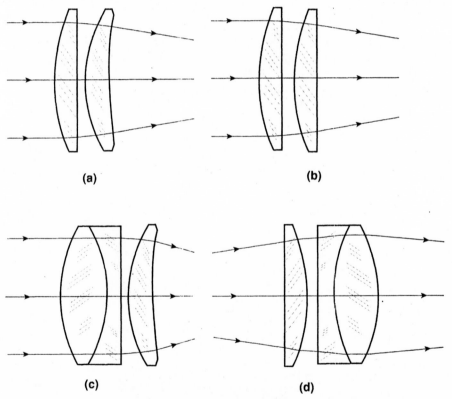

(a)   (b)

(c)   (d)

Figure 6.21 When lenses are used at high speed (large NA or small f-number), spherical aberration is the usual problem. It can be reduced by using two elements instead of one. The first should be oriented to minimize spherical for the object location and the second shaped for *its* object location. For distant objects the best arrangements are shown in (A), (B), and (C). The first element is planoconvex and the best shape for the second is meniscus; if the second is also a planoconvex it should be oriented as in (B). If one element is a doublet, it should face the longer conjugate as indicated in (C) and (D).

tant object, followed by the singlet as in Fig. 6.21C. If both are doublets, put their strongly curved surfaces toward the object. If the application is microscope like, then of course the arrangement is reversed as shown in Fig. 6.21D.

When the system is to work at finite conjugates, for example, at one-to-one or at a small magnification, then the best arrangement is usually with the convex surfaces facing each other (provided that the angular field is small). Figure 6.22 shows pairs of singlets and doublets working at one-to-one. This arrangement allows each half of the combination to work close to its design configuration, i.e., with the object at infinity, and if the angular field is not large, the space may be varied to get a desired track length. If the magnification is not one-to-one, using different focal lengths (whose ratio equals the magnification) can be beneficial.

A *projection condenser* is usually two or three elements, shaped and arranged to minimize spherical aberration. There is often an aspheric surface. With two elements, the more strongly curved surfaces face each other; if they are not the same power, the stronger (shorter focal length) faces the lamp. With three elements, the one nearest the lamp is often meniscus with the concave surface facing the lamp. The other

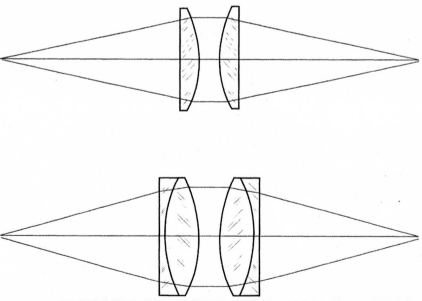

Figure 6.22 For small fields and magnifications which are close to 1:1, the lenses should be oriented facing each other as shown in order to minimize the spherical. At 1:1 the light is collimated between the lenses; at other magnifications it is nearly so.

two are often planoconvex, with their curved sides facing. If the elements are spherical-surfaced, they should each be approximately the same power. If one is aspheric, it is often stronger than the others and is the one next to the lamp.

Eyepieces and magnifiers. Very good magnifiers can be made from two planoconvex elements with the curved sides facing each other as shown in Fig. 6.23A. This arrangement works well, either close to the

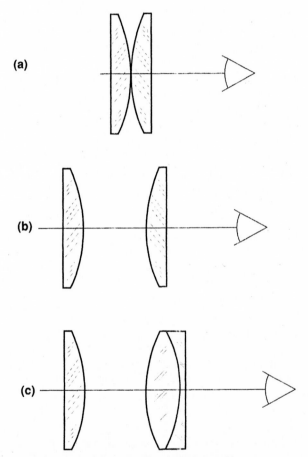

Figure 6.23 (*A*) Two planoconvex elements, convex to convex, make a good magnifier which works well both near the eye and at a distance. (*B*) For use as a telescope eyepiece, the spacing is increased to reduce the coma and lateral color (and to allow the left-hand lens to act as a field lens). (*C*) The Kellner eyepiece uses a doublet as the eyelens to further correct the lateral color. This eyepiece is often found in ordinary prism binoculars.

eye or at arm's length. As a telescope eyepiece, the spacing between them is often increased to about 50 or 75 percent of the singlet focal length, so that one element acts as a field lens, as in Fig. 6.23B; this increased spacing also reduces the lateral color and helps with coma and astigmatism. (If the elements have different focal lengths, the lens near the eye should have the shorter focal length.) This is the classical "Ramsden" eyepiece. If the eyelens is a doublet, it is the "Kellner" eyepiece shown in Fig. 6.23C; usually the flatter side of the doublet faces the eye. Some versions of this popular binocular eyepiece are closely spaced, and some are used in a reversed orientation. A few trials with a graph paper target and your eye at the exit pupil location will tell you which arrangement suits your stock lenses the best.

Two achromats work even better. With the strong curves of two identical achromats facing each other as in Fig. 6.24A, this makes one of the best general-purpose magnifiers and eyepieces. This is the "Ploessl" or "symmetrical" eyepiece, justly popular for its high quality, low cost, versatility, and long eye relief. Depending on exactly what the shape of your doublet is and what your eye relief is, you may want to reverse the orientation of one or the other (but not both) of the doublets as indicated in Fig. 6.24B and C.

Wide-field combinations. Let's face it right up front. It's *very* difficult to put together stock elements so that they perform well over a wide field of view. Usually your best bet is to obtain a corrected assembly such as a triplet anastigmat or a camera lens. But there are a few things we can do to optimize the situation when we don't have a suitable anastigmat available.

As mentioned earlier, to obtain a wide-field coverage we often must sacrifice the image quality in the center of the field. We have two basic tools which we can use to improve the image quality at the edge of the field. One is the placement of the aperture stop, and the other is the *symmetrical principle.* If a system is symmetrical about the stop (in a left-to-right sense as shown in Fig. 6.25), then the system is free of coma, distortion, and lateral color. Strictly speaking, the system must work at unit magnification to be fully symmetrical, but much of the benefit of symmetry is obtained even if the object is at infinity. Of course, symmetry works whether we're doing wide or narrow fields of view. But whereas we orient the elements "strong-side-facing" as in Fig. 6.25A to get the least spherical aberration in a narrow-field application, we usually want the strong surfaces facing outward for wider fields of view as in Fig. 6.25B. Planoconvex or meniscus elements are the shapes of choice for this. The elements are

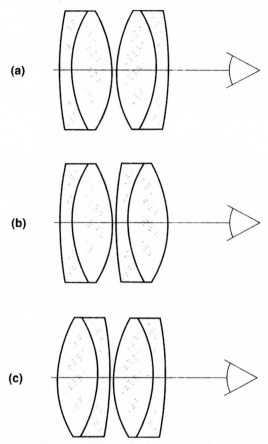

(a)

(b)

(c)

Figure 6.24 (*A*) Two doublets, crown to crown, make
an excellent eyepiece and also an excellent magnifier. This is the symmetrical, or Ploessl, eyepiece. (*B*)
and (*C*) Depending on the shape of the doublets and
the eye relief of the telescope, one of these alternate
orientations may work well as an eyepiece.

spaced a modest, but significant, distance from the aperture stop,
which is midway between them. The spacing is significant because it
affects the astigmatism; there is an optimum spacing which yields the
best compromise between the amount of astigmatism and the flatness
of the field.

Relay systems. For a relay system which requires some given magnification, consider using two achromats, such that the ratio of their
focal lengths equals the desired magnification, and the sum of their
focal lengths is approximately equal to the desired object-to-image

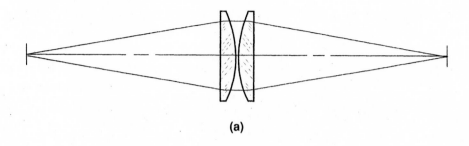

(a)

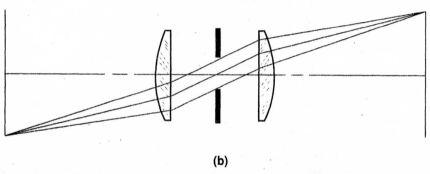

(b)

Figure 6.25 Left to right (or mirror) symmetry will automatically eliminate coma, distortion, and lateral color. With two planoconvex elements, the orientation shown in (A) would be best for a small field, but for a wider field, the orientation in (B) will usually work better.

distance, as shown in Fig. 6.26. The rays in the space between the lenses will be collimated, and the spacing between them will not be a critical dimension. Note that a 45° tilted-plate beam splitter can be used in a collimated beam without introducing astigmatism. If the achromats are corrected for an infinite object distance, the relay image will also be corrected.

The two-achromat relay can produce an excellent image over a small field. A wider-field system can be made from two photographic lenses, again used face to face, with collimated light between them, as shown in Fig. 6.27. Since photo lenses are longer than the achromatic doublets we discussed in the preceding paragraph, one must be aware of vignetting. Most photographic objectives vignette when used at full aperture, often by as much as 50 percent. For an oblique beam (tilting upward as it goes left to right) the beam is clipped at the bottom by the aperture of the left lens, and clipped at the top by the right-hand lens. When two camera lenses are used face to face, their vignetting

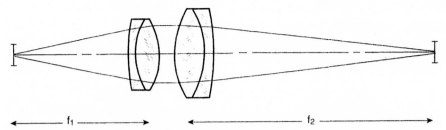

f_1 ← → ← f_2 →

Figure 6.26 A well-corrected narrow-field relay can be made from two achromatic doublets by choosing their focal lengths so that their ratio equals the desired magnification $m = -f_2/f_1$. When this is done, the light between the lenses is collimated and each lens works at its design conjugates (assuming the lenses were designed for an infinitely distant object).

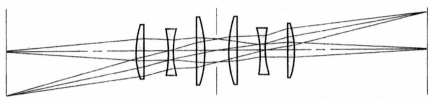

Figure 6.27 When a wider field than two doublets (as shown in Fig. 6.26) can cover is needed, two camera lenses can be used, face to face, to make a high-quality relay system. If the relay is to have magnification, the focal lengths of the lenses should be chosen so that their ratio equals the magnification. If an iris diaphragm is to be used, it should be located between the lenses (unless the field is small). Note that with some lenses vignetting may be a problem.

characteristics are usually such that the combination has much worse vignetting than either lens alone. Thus this sort of relay is usually limited by vignetting to a smaller field than one might expect. Note also that if an iris diaphragm is to be used, it should be between the lenses, rather than using the iris of one of the lenses (unless the field of view is quite small). This arrangement of readily available stock camera or enlarging lenses makes an excellent, well-corrected finite conjugate imaging system.

The above technique of using photo lenses has the virtue of using them as they were designed to be used, namely, with one conjugate at infinity. Most photo lenses retain their image quality down to object distances of about 25 times their focal lengths, more or less, depending on the design type. But at close distances the image quality deteriorates. In general, high-speed lenses tend to be quite sensitive to object distance. Slower (i.e., low NA, large f-number) lenses can be used successfully over a wider range of conjugate distances.

A *close-up attachment* is simply a weak positive lens placed in front of a camera lens. If, as shown in Fig. 6.28, the focal length of the

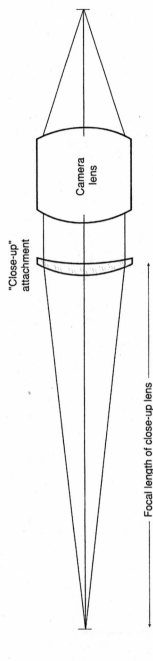

"Close-up" attachment

Camera lens

←——————— Focal length of close-up lens ———————→

Figure 6.28 Many camera lenses lose image quality when the object is close. A "close-up attachment" is simply a weak positive element whose focal length approximates the object distance, so that the light is collimated for the camera lens. The attachment is usually a meniscus lens whose shape is a compromise between minimum spherical aberration and minimum coma and astigmatism.

attachment lens is approximately equal to the object distance, then the object is collimated (imaged at infinity) and the camera lens sees the object as if it were at infinity. The attachment lens is ideally a meniscus, with the concave side facing the camera lens (so that it wraps around the stop), although a planoconvex form is often quite acceptable. If the field is quite narrow, the reverse orientation of the lens *might* be better. Note that the use of a close-up attachment is equivalent to combining two positive lenses to get a lens with a shorter focal length. You can also use a weak negative focal length attachment to increase the focal length of a camera lens which is too short for your application.

Beam expander.   A laser beam expander is simply a telescope used "backward" to increase the diameter and to reduce the divergence of the laser beam. The galilean form of telescope is the most frequently used because it can be executed with simple elements and has no internal focus point (which might induce atmospheric breakdown with a high-power laser). The Kepler telescope can also be used, and its internal focal point can provide a spatial filter capability, but it is more difficult to correct the Kepler because both components are positive, converging lenses.

Since the laser light is monochromatic and the beam angle is small, we are mostly concerned with correcting spherical aberration. The objective (positive) component of the galilean is the big contributor of spherical aberration, so it is important that it be shaped to minimize spherical. If the expander is to be made from two simple elements as diagramed in Fig. 6.29A, the negative element must contribute enough overcorrected spherical to balance that from the objective lens. Thus our "stock lens" choice is often a planoconvex element for the objective and a planoconcave element for the negative, with both lenses oriented so that their plano sides face the laser. (A meniscus form for the negative has more overcorrected spherical and might produce a better correction.) For higher-power beam expanders, a well-corrected doublet objective is necessary; it should be combined with a planoconcave element with its concave side facing the laser as shown in Fig. 6.29B to minimize its overcorrection of the spherical aberration.

6.11   Sources

Sources of stock lenses

The following are some of the companies which stock lenses. Most have catalogs. Some have their catalogs on disk; many of these include the lens prescriptions (radii, thickness, index), and a few have

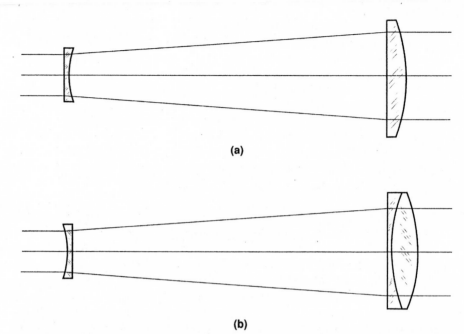

(a)

(b)

Figure 6.29 A low-power laser beam expander can be made from a planoconvex "eye-lens" and a planoconvex "objective," with both plano sides facing the laser. For higher powers an achromatic doublet is used as the objective to reduce the spherical aberration, and the planoconcave negative element is reversed.

free computer programs which can be used to calculate the performance of their lenses.

Ealing Electro-Optics, Inc.
89 Doug Brown Way
Holliston, MA 01746
Tel: 508/429-8370;
Fax: 508/429-7893;
http://www.ealing.com

Edmund Scientific
101 East Gloucester Pike
Barrington, NJ 08007
Tel: 609/573-6852;
Fax: 609/573-6233;
John_Stack@edsci.com

Fresnel Optics, Inc.
1300 Mt. Read Blvd.
Rochester, NY 14606

Tel: 716/647-1140;
Fax: 716/254-4940

Germanow-Simon Corp., Plastic
  Optics Div.
408 St. Paul St.
Rochester, NY 14605-1734
Tel: 800/252-5335;
Fax: 716/232-2314;
gs optics@aol.com

Janos Technology Inc.
HCR#33, Box 25, Route 35
Townshend, VT 05353-7702
Tel: 802/365-7714;
Fax: 802/365-4596;
optics@sover.net

JML Optical Industries, Inc.
690 Portland Ave., Rochester, NY
14621-5196
Tel: 716/342-9482;
Fax: 716/342-6125;
marty@jmlopt.com
http://www.jmlopt.com

Melles Griot, Inc.
19 Midstate Drive, Ste. 200
Auburn, MA 01501
Tel: 508/832-3282;
Fax: 508/832-0390;
76245,2764@compuserve.com

Newport Corporation
1791 Deere Ave., Irvine, CA 92714
Tel: 714/253-1469;
Fax: 714/253-1650;
pgriffith@newport.com

Optics for Research
P.O. Box 82, Caldwell, NJ
07006-0082
Tel: 201/228-4480;
Fax: 201/228-0915;
dwilson@ofr.com

Optometrics USA, Inc.
Nemco Way, Stony Brook Ind. Park
Ayer, MA 01432
Tel: 508/772-1700;
Fax: 508/772-0017;
opto@optometrics.com

OptoSigma Corp.
2001 Deere Ave.
Santa Ana, CA 92705
Tel: 714/851-5881;

Fax: 714/851-5058;
optosigm@ix.netcom.com

Oriel Instruments
250 Long Beach Blvd., P.O. Box 872
Stratford, CT 06497-0872
Tel: 203/377-8282;
Fax: 203/378-2457;
res_sales@oriel.com

Reynard Corporation
1020 Calle Sombra
San Clemente, CA 92673
Tel: 714/366-8866;
Fax: 714/498-9528

Rodenstock Precision Optics, Inc.
4845 Colt Road, Rockford, IL
61109-2611

Rolyn Optics
706 Arrowgrand Circle, Covina, CA
91722-9959
Tel: 818/915-5707;
Fax: 818/915-1379

Spectral Systems
35 Corporate Park Drive
Hopewell Junction, NY 12533
Tel: 914/896-2200;
Fax: 914/896-2203

Spindler & Hoyer Inc.
459 Fortune Blvd.
Milford, MA 01757
Tel: 508/478-6200;
800/334-5678;
Fax: 508/478-5980

Their catalog on disk includes an
"Optical Design Program for WIN-
DOWS."

Optical design programs

At least two full-featured optical design programs are available free
by downloading from the internet. There may be others. The two that
I currently know of are:

"KDP, the Free Optical Design
 Program"
Engineering Calculations
1377 E. Windsor Rd. #317
Glendale, CA 91205
Tel & Fax: 818/507-5705
email: kdpoptics@themall.net
"Available via anonymous FTP at
www.kdpoptics.com in
directory/users/kdpoptics"

"OSLO LT"
Sinclair Optics, 6780 Palmyra Road
Fairport, NY 14450
Tel: 716/425-4380;
Fax: 716/425-4382
email: oslo@sinopt.com
Web site URL http://www.sinopt.com

("Visit our home page. Click "OSLO
LT." Download your free copy.") This
is the program OSLO LITE except
with "no file save, hard copy by
screen capture only." The program
has some 3000 lenses included, with
prescriptions.

Directories

Several directories are available which can help in locating sources of
optical things. Probably the most complete is the *Photonics Buyer's
Guide,* published by Laurin Publishing Co., Inc., Berkshire Common,
P.O. Box 4949, Pittsfield, MA 01202-4949, Tel:413/499-0514,
Fax:413/442-3180, email: Photonics@MCIMail.com. This is the lead
volume of a four-volume set; it lists optical products by category, giving sources for each type of product. A second volume, the *Photonics
Corporate Guide,* lists the names, addresses, etc., of the source companies. *Laser Focus World* magazine and *Lasers & Optronics* magazine also publish optical buyers guides which are distributed to sub-
scribers.

Bibliography

General optical engineering books

Habell and Cox, *Engineering Optics,* 1948, Pitman (the optics of measurement and alignment).

Hardy and Perrin, *The Principles of Optics,* 1932, McGraw-Hill (despite its age, it is a sound and wide-ranging coverage of optics).

Jacobs, *Fundamentals of Optical Engineering,* 1943, McGraw-Hill (basic optical engineering with emphasis on military optical instruments circa 1943).

Kingslake, *Optical System Design,* 1983, Academic (excellent coverage of the topic; describes the principles of a tremendous variety of optical systems).

Kingslake et al., *Applied Optics and Optical Engineering,* 11 vols, 1965 to 1992, Academic Edited by Kingslake, Thompson, Shannon, Wyant (a complete library of optical engineering in eleven volumes with contributions from about a hundred authorities).

Kissam, *Optical Tooling,* 1962, McGraw-Hill (the optics of alignment and measurement).

Smith, *Modern Optical Engineering: The Design of Optical Systems,* 2d ed., 1990, McGraw-Hill (perhaps the most widely used optical engineering text; both a classroom text and self-study manual plus a convenient reference).

Walker, *Optical Engineering Fundamentals,* 1995, McGraw-Hill (a gentle introduction to the basics of optical engineering).

Books on lens design

Conrady, *Applied Optics and Optical Design,* part 1, 1929/1957, Dover; part 2, 1960, Dover (based on the use of logarithms for calculation; presents a very sound, classical understanding of lens design principles).

Cox, *A System of Optical Design,* 1964, Focal (a complete approach to lens design; includes prescriptions and analysis of some 300 lens design patents).

Hopkins and Hanau, *Military Handbook 141, Optical Design,* 1962, Defense Supply Agency (republished by Sinclair Optics, Fairport, NY) (authoritative contributions by many authors on a wide range of topics related to optical design).

Kingslake, *Lens Design Fundamentals,* 1978, Academic (presents a sound basic approach; based on calculation with an electronic pocket calculator; written by the master).

Laikin, *Lens Design,* 1991, Dekker (includes about 100 prescriptions with computed MTF analysis).

Riedl, *Optical Design Fundamentals for Infrared Systems,* 1995, SPIE (basic, clear instruction for IR design; especially good for preliminary design steps).

Smith, *Modern Lens Design*, 1992, McGraw-Hill (a companion volume to Smith's *Modern Optical Engineering*; instruction in both basic and advanced lens design, plus almost 300 lens design prescriptions with a complete raytrace analysis of each).

Other books on optics

Two monthly magazines devoted to optics, that are available free of charge to qualified individuals and are well worth reading are:

Photonics Spectra, Berkshire Common, P.O. Box 4949, Pittsfield, MA 01202-4949, 413/499-0514.

Laser Focus World, 10 Tara Blvd., Fifth Floor, Nashua, NH 03062, 603/0123.

Accetta (ed.), *The Infrared and Electro-Optical Systems Handbook*, 8 vols. 1993, SPIE Press.

Allard, *Fiber Optics Handbook*, McGraw-Hill.

Bass (ed.), *Handbook of Optics*, 2 vols., 1995, McGraw-Hill.

Born and Wolf, *Principles of Optics*, 1980, Pergamon.

Brown, *Modern Optics*, 1965, Reinhold.

Cselt, *Fiber Optic Communications Handbook*, McGraw-Hill.

Driscoll (ed.), *Handbook of Optics*, 1978, McGraw-Hill.

Elliott, *Microlithography*, 1986, McGraw-Hill.

Gaskill, *Linear Systems, Fourier Transforms, and Optics*, 1978, Wiley.

Goodman, *Introduction to Fourier Optics*, 1968, McGraw-Hill.

Hecht and Zajac, *Optics*, 1974, Addison-Wesley.

Hecht, *The Laser Guidebook*, McGraw-Hill.

Jenkins and White, *Fundamentals of Optics*, 1976, McGraw-Hill.

Kao, *Optical Fiber Systems*, McGraw-Hill.

Keiser, *Optical Fiber Communications*, McGraw-Hill.

Kingslake, *A History of the Photographic Lens*, 1989, Academic.

Kingslake, *Optics in Photography*, 1992, SPIE Press.

Levi, *Applied Optics*, 1968, Wiley.

Macleod, *Thin Film Optical Filters*, McGraw-Hill.

Malacara, *Optical Shop Testing*, 1978, Wiley.

Melzer and Moffit, *Head Mounted Displays*, 1997, McGraw-Hill.

Meyer-Arendt, *Introduction to Classical and Modern Optics*, 1972, Prentice-Hall.

Miller and Friedman, *Photonics Rules of Thumb*, 1996, McGraw-Hill.

Rancourt, *Optical Thin Film Users' Handbook*, McGraw-Hill.

Sears, *Optics*, 1949, Addison-Wesley.

Sibley, *Optical Communications*, McGraw-Hill.

Siegman, *Introduction to Lasers and Masers*, 1971, McGraw-Hill.

Southall, *Mirrors, Prisms and Lenses*, 1933, Macmillan, 1964, Dover.

Stover, *Optical Scattering*, McGraw-Hill.

Strong, *Concepts of Classical Optics*, 1958, Freeman.

Strong, *Procedures in Applied Optics*, 1989, Dekker.

Syms and Cozens, *Optical Guided Waves and Devices*, McGraw-Hill.

Thelen, *Design of Optical Interference Coatings*, McGraw-Hill.

Walsh, *Photometry*, 1958, Constable, 1965, Dover.

Wayant and Ediger (eds.), *Electro-Optics Handbook*, 1994, McGraw-Hill.

Welford, *Aberrations of the Symmetrical Optical System*, 1974, Academic.

Welford, *Aberrations of Optical Systems*, 1986, Adam Hilger.

Wolfe and Zissis (eds.), *The Infrared Handbook*, 1978/1985, ERIM.

Wyatt, *Electro-Optical System Design*, McGraw-Hill.

Yoder, *Mounting Lenses in Optical Instruments*, 1995, SPIE.

Yoder, *Opto-Mechanical Systems Design*, 1986, Dekker.

Index

Aberration, chromatic, 143, 144, 148
Aberrations, 163–171
  vs. lens shape, 167–169
  Seidel, 163
Achromatic doublet, 144
  capabilities, 172, 173
Aerial image modulation, 122
Afocal attachment, 66–68
  cylindrical, 83, 84
  prismatic, 85–87
Afocal magnification, 125
Afocal mirror systems, 94, 95
Afocal system, 57–68
AIM curve, 121–124
Air-equivalent distance, 88, 91
Airy disk, 117–119
Algebraic layout solution, 131–135
Alignment of mock-up, 160
Amtir, 149, 150
Anamorphic system, 82–87
Angular magnification, 57–59, 125
Angulon, 173
Aperture capability, 171–173
Aperture ratio of LCD, 108, 109
Aperture stop, 49, 50, 54, 64, 178
Aperture, numerical (NA), 54
Aperture, relative, 54
Apertures, decentered, 95, 96
Apparent field, 58, 62
Arc lamp, 104
Array, beam homogenizer, 106, 107
Aspheric reflector, deep, 106
Aspheric surface, 174
Assumptions, 1
Astigmatism, 163–166, 185

Astronomical telescope, 59–61
Athermalization, 144, 145, 150
Attachment, afocal, 66–68
Automatic lens design, layout by, 138–140
Axial chromatic aberration, 143, 144, 148, 163–166
Axial ray, 26
Axis, optical, 1
Axis, projection, 36

Back focal length, 4
  calculation of, 20
  measurement of, 155–157
Barrel distortion, 165
Beam diameter, 126
Beam expander, 65, 66, 189, 190
Beam homogenizer array, 106, 107
Beam intensity of conic reflectors, 105
Beam obscuration, 94
Beam splitter, 80, 186
Bibliography, 193, 194
Biconvex lens usage, 176–180
Binary surface, 175
Binocular optics, 38, 41, 60
Black flocked paper, 162
Blackbody, 113
Books, 193, 194
Bravais system, 68–70
Bravais, anamorphic, 84, 85
Brightness, 110–115
Buyer's guides, 192

Camera lens capabilities, 172

Camera lenses as relays, 186, 187
Candle (candela), 113
Capabilities, lens types, 171–173
Cardinal points, 2, 3
  calculation of, 20
  location, 5
Cassegrain mirror system, 34, 35, 92, 93, 173
Catalog lenses, 153–192
Catalogs, 189–191
CDM (chromatic difference of magnification), 143
Cemented doubled usage, 179–182
Chief ray, 26, 63
Chromatic aberration, 143, 144, 148, 163–166
Chromatic difference of magnification (CDM), 143
Close-up attachment, 187–189
Coherent illumination, 121
Cold stop, 145–151
Collimated light, 68
Collimator, 82
  LED, 86, 87
Coma, 163–166, 184
  of conicoid, 106
Combination of two components, 27–35
Combinations of stock lenses, 180–189
Combiner mirror, 80
Component raytracing, 23–26
Component, definition, 2
Components, combination of two, 27–35
Compound microscope, 78, 79
Computer code, layout by, 138–140
Concept, thin lens, 22
Condenser, 174
Condenser, projection, 101–108, 182, 193
Confocal paraboloids, 94, 95
Conic reflectors, beam intensity, 105
Conic section, coma of, 106
Conjugates, 8
Conservation of brightness, 110
Constant deviation, 39, 42
Constant magnification, 59
Constrained targets, 140
Conventions, 1, 2
Converging lens, 3
Copy machines, 175
Corner cube, 39
Cosine fourth, 112
Critical angle, 38

Critical illumination, 104, 105, 110
Cube corner, 39
Curvature of field, 163–171
Curvature, 2
Cutoff frequency, 120
Cylinder lenses, 82–85
  alignment, 160

Damped least squares (DLS), 139
Dawes criterion for resolution, 119
Decentered apertures, 95, 96
Depth of field, 127
Depth of focus, 126–128
Derotator prism, 39, 42
Detector and source, interchangeability, 101
  size limit, 126
  smallest possible, 102, 103
Diameter determination, 142, 148
Diamond turned surface, 174
Differential adjustment, 135–138
Diffraction, limit, 117–125
  limited, 118, 119, 121
  vs. sensor limits, 123–125
Diffractive microlenses, 108
Diffractive surface, 175
Digital micromirror device (DMD), 106, 110
Diopter, 4, 18
Directories, 192
Distance, sign convention, 2
Distortion, 163–166, 184
DLS (damped least squares), 139
DMD (digital micromirror device), 106, 110
Double Gauss, 173
Doublet lens capabilities, 172, 173
Doublet usage, 179–182

Ealing Electro-Optics, 190
Edmund Scientific, 162, 190
Effective focal length (efl), 3
  calculation of, 20
  measurement of, 157–159
Element powers for achromatism, 144
Element, definition, 2
Ellipsoid mirror, 104, 106
Empty magnification, 124, 125
Endoscope, 175

Engineering Calculations, 192
Enlarging lenses, 172
Entrance pupil, 51–58
Entrance window, 57
Equations, worked out ray, 132–135
Equiconvex lens usage, 176–180
Equivalent air distance, 34
Erecting prism systems, 38, 41
Etendue, 26
Exit pupil, 51–58, 64, 146, 162
Exit window, 57
Expander, beam, 65, 66, 189, 190
Eye relief, 55, 56, 71–73
Eyepiece, 178, 183–185
    testing, 161

F-number, 54, 142
    working, 55
Fiber optic endoscope, 176
Field coverage, 171–173
Field curvature, 163–171
Field flattener lens, 170, 171
Field glasses, 61, 62
Field lens, 70–75
    fresnel, 175
Field of view, 58, 59, 62, 126, 142
Field stop, 57, 64
Field, depth of, 127
Field, real and apparent, 58, 62
Fisheye lens, 81, 173
Flat black paint, 162
Flocked paper, 162
Floquil flat black paint, 162
Focal length, 3, 4
    back, 4
    calculated, 17
    calculation of, 20
    front, 4
    measurement of, 157–159
Focal points, 2, 3
Focus shift, thermal, 148–151
Focus, depth of, 126–128
Focusing attachment for anamorphics, 84, 85
Folding a prism system, 37
Foot-lambert, 113
Footcandle, 113
Four-component ray equations, 132–134
Free variable, 135
Frequency, cutoff, 120

Fresnel lens, 106, 108, 174
Fresnel Optics Inc, 190
Front focal length, 4

Gain, screen, 115
Galilean beam expander, 189, 190
Galilean telescope, 61, 62
Gauss points, 2, 3
Gauss's equation, 8
Gaussiam beam, 121
Germanium, 149, 150
Germanow-Simon Corp, 190
Glare stop, 63, 64, 146, 162
Gradient index, 175, 176
Gregorian mirror system, 94
Grin rod, 175
Guide, buyer's, 192

Head mounted display (HMD), 79, 80, 178
Head-up display (HUD), 79, 80, 178
Height, image, 10
Height, ray, 17
Helmet mounted display, 79, 80
High speed applications, 180–182
HMD (head mounted display), 79, 80, 178
HUD (head-up display), 79, 80, 178
Hyperfocal distance, 127

Illumination, 110–115
    critical, 104, 105, 110
    Koehler, 101–104
    smoothing, 105–107
Image brightness, 110
    equations, 4–12
    height, 10
    motion, 12–15
    orientation, 38
    plane, tilted, 35–37
    real or virtual, 6
    shift by glass plate, 38
Index, effect of on curvatures, 141
    effect on spherical aberration, 168
    sign of, 2
    thermal coefficient, 144
Infrared system layout, 145–151
Integrator bar, 105
Intensity, 113
Interchangeability, source and detector, 101

Intersection length, ray, 17
Invariant, 26, 66, 125, 126
Inversion prisms, 38, 42
Irradiance, 110–115

Janos Technology, 190
JML Optical Ind., 191

KDP optical design program, 192
Kellner eyepiece, 183, 184
Kepler beam expander, 189
Kepler telescope, 59–61
    layout of, 135, 136
Keystone distortion, 36, 37
Kinoform surface, 175
Koehler illumination, 101–104

Lagrange invariant, 26, 66, 125, 126
Lambert's law, 111
Lambert, 113
Landscape lens, 172, 173
Laser beam expander, 189, 190
Laser beam, 121
Laser Focus World, 192, 194
Lasers & Optronics, 192
Lateral color, 143, 144, 163–166, 184
Laurin Publishing Co., 192
Layout by computer code, 138–140
Layout of a lens erecting telescope,
    136–138
Layout of a system, how to, 129–151
Layout of Kepler telescope, 135, 136
Layout steps, 129
LCD (liquid crystal display), 106–109
LED collimator, 86, 87
Lens bench, simple, 159–161
Lens design books, 193, 194
Lens shape vs. aberration, 167–169
Lens types, capabilities of, 173
    converging, 3
    field, 70–75
    relay, 70–75
Lens-erecting telescope, 63–65
    layout of, 136–138
Light pipe, 105
Limits, system, 117–128
Line resolution, 120
Liquid crystal display (LCD), 106–109

Local optimum, 139
Longer system, 131
Longitudinal magnification, 10–12
Luminance, 110–115
Lux, 113

Magazines, 194
Magnification, 125
    angular, 57–59
    Bravais system, 68–70
    constant, 59
    empty, 124, 125
    equations, 4–12
    longitudinal, 10–12
    magnifier, 76–78
    measurement of, 157–159
    microscope, 76–79
    transverse, 6, 8
Magnifier, 76–78, 183–185
    testing, 161
Magnifying glass, 178, 179
Magnifying power, 57, 58
Measurements, 155–159
Melles Griot, Inc, 191
Meniscus doublet, 180
Merit function for layout solutions,
    138–140
Meter-candle, 113
Microlens array for LCD, 108, 109
Microscope, 76–79
    compound, 78, 79
    objectives, 172
Minimized targets, 140
Mirror system, 34, 35, 88–96
Mirror, cardinal points, 5
Mirrors, 37–40
Mock-up of system, 159–162
Modulation, 120
Modulation transfer function (MTF),
    120–124
Motion, image, 12–15
Motion, object, 12–15
MTF (modulation transfer function),
    120–124
Multiconfiguration feature for layout, 139

NA (numerical aperture), 54, 55
Newport Corp, 191
Newton's equation, 4–6

Nodal points, 10, 12, 15
Numerical aperture (NA), 54, 55
Numerical layout solution, 135–138

Object, contrast for resolution, 123
    distance effect, 187
    distance vs. aberration, 167, 169
    distance vs. performance, 172
    motion, 12–15
    plane, tilted, 35–37
Obscuration in mirror systems, 94
Off-axis parabola, 96
Opera glasses, 61, 62
Optical axis, 1
Optical design programs, 191, 192
Optical engineering books, 193
Optical invariant, 66, 125, 126
Optical software, 191, 192
Optical systems, basic, 49–99
Optics for Research, 191
Optimization in computer codes, 139
Optimum configuration, 131
Optimum, local, 139
Optometrics USA, 191
OptoSigma Corp, 191
Oriel Instruments, 191
Orientation of image, 38
OSLO LT, 192
Overconstrained problem, 140
Overhead projector, 175

Paint, flat black, 162
Parabola, off-axis, 96
Paraboloid reflector, 106
Paraboloids, confocal, 94, 95
Paraxial ray, linearity, 26, 37
    scaling, 27
    slope, 2
Paraxial raytracing, 17–22
Paraxial, meaning of, 22
Passive thermal compensation, 145–151
Performance limits, 117–128
Periscope, 73–75
Petzval field curvature, 163–171, 173
Phot, 113
Photographic lenses as relays, 186, 187
Photometry, 110–115
Photonics Buyer's Guide, 192
Photonics Corporate Guide, 192

Photonics Spectra, 194
Pincushion distortion, 165
Pinhole target, 160, 161
Plane, principal, 3
Planoconvex lens usage, 176–180
Plastic fresnel lenses, 174
Ploessl eyepiece, 184, 185
Point resolution, 118
Points, cardinal and Gauss, 2, 3
Points, focal, 2, 3
Points, nodal, 10, 12, 15
Points, principal, 2, 3
Porro prisms, 38, 41
Power changer, 67
Power emitted, 110
Power, 4
    achromatic doublet elements, 144
    change with temperature, 145
    magnifying, 57, 58
    surface, 18
Primary aberrations, 163
Principal plane, 3
Principla points, 2, 3, 15
Principal ray, 26, 63
Principal surface, 2
Prism anamorphic, 85–87
Prisms, 37–40
Projection axis, 36
Projection condenser, 101–108, 174, 182, 183
Projection lens, 101–104, 107, 108
Projector, motion picture, 104
Pupil, 49–57, 63, 125
    afocal, 58
    decentration, 121
    entrance, 51–58
    exit, 51–58, 64

Quarter-wave, resolution limit, 121
    depth of focus, 127, 128

Radiance, 110–115
Radiometer illumination, 102, 103
Radiometry, 110–115
Radius, sign of, 2
Ramsden eyepiece, 184
Ray:
    axial, 26
    chief, 26

Ray (*Cont.*):
equations, worked out, 132–135
height, 17
intersection length, 17
principal, 26
sketching, 6, 7, 12, 15, 16
slope, 17
(*See also* Paraxial ray)
Rayleigh criterion for resolution, 118–120
Rayleigh depth of focus, 127, 128
Rayleigh limit for aberration, 121
Raytracing:
components, 23–26
equations, 23
mirror, 34, 35
paraxial, 17–22
thin lens, 23–26
Real field, 58, 62
Real image, 6
Red wax, 159, 160
Reflecting systems, 88–96
Reflection, sign of index after, 2
Reflection, total internal (TIR), 38
Reflectors, 37–40
Refraction, 2
Refractive index, sign of, 2
Relative aperture, 54, 142
capability, 171–173
Relay combinations, 182, 184, 186, 187
Relay lens, 70–75, 177
Relay systems, 185–187
Relief, eye, 55, 56
Resolution, 118–125
limit, 123
of eye, 121
testing, 160–162
Retrodirector, 39
Retrofocus, 80, 81, 173
Reversed telephoto, 80, 81, 173
Reynard Corp, 191
Riflescope, 124
Rodenstock Precision Optics, 191
Rolyn Optics, 191
Rough sketch of elements, 140–143

Salvage lenses, 154, 155
Scaling paraxial rays, 26, 27
Scheimpflug condition, 16, 35–37
Schmidt system, 173
Schwarzschild mirror system, 92, 93

Screen brightness, 115
Screen gain, 115
Seidel aberrations, 163
Selfoc rods, 175
Semicoherent illumination, 121
Sensor limits, 121–125
Sensor vs. diffraction limits, 123–125
Separated components, 27–35
Sign convention, ray slope, 18
Sign conventions, 2
Silicon, 148
Simple lens, cardinal points, 5
Sinclair Optical, 192
Singlet usage, 176–179
Size limit, source and detector, 126
Sketching lens elements, 140–143
Sketching rays, 6, 7, 12, 15, 16
Slide projector condenser, 103, 104
light output, 113–115
Slide viewer, 77, 78
Slope, paraxial ray, 2, 17
Smallest possible detector, 102, 103
Smallest possible source, 102
Smoothing, illumination, 105–107
Snell's law, 2
Software, lens prescriptions, 154
optical design, 191, 192
Solid angle, 112
Solution, algebraic, layout, 131–135
numerical, layout, 135–138
Solves for ray slopes and heights, 139
Source and detector interchangeability, 101
Source, size limit, 126
smallest possible, 102
Spacing, sign of, 2
Sparrow criterion for resolution, 119, 120
Spatial filter, 189
Spectral Systems, 191
Spherical aberration, 156, 163–170
in beam expander, 189, 190
reduction, 180–182
Spindler & Hoyer, 191
Splitting elements, 142, 167, 170
Stilb, 113
Stock lenses, 153–192
combinations, 180–189
drawbacks, 153–155
sources, 189–191
Stop, aperture, 49, 50, 54, 64
position vs. aberration, 167

Stop, aperture (*Cont.*):
  cold, 145–151
  field, 57, 64
  glare, 63, 64, 162
  telecentric, 57
Stops, 49–57
Stray light, 156, 161, 162
Surface power, 18
Symmetrical eyepiece, 184, 185
Symmetrical principle, 184, 185
System characteristics, 129, 130
System layout, how to, 129–151
System length requirements, 132

Tacky wax, 159, 160
Tapered light pipe, 105
Target, resolution, 160–162
Targets for merit function, 138–140
Telecentric system, 57
Telephoto attachment, 66–68
Telphoto system, 80, 81
Telescope objective, 180
  capabilities, 172, 173
Telescopes, 59–65
  astronomical, 59–61
  galilean, 61, 62
  Kepler, 59–61
  lens-erecting, 63–65
  terrestrial, 63–65
Terrestrial telescope, 63–65
Tessar, 173
Thermal expansion coefficient, 144
Thermal focus shift, 148–151
Thermal index coefficient, 144
Thin lens, 22
  raytracing, 23–26
Third ray creation, 27
Three-component ray equations, 132–134

Threshold curve, sensor, 121–123
Throughput, 26
  limits, 117–128
Tilted image plane, 35–37
Tilted object plane, 35–37
Total internal reflection (TIR), 38
Transit, 124
Transverse magnification, 6, 8
Triplet, 173
Tunnel diagram, 40
Two-component combinations, 27–35
Two-component ray equations, 132–134
Two-mirror systems, 91–95

Underconstrained problem, 139, 140
Unfolding a prism system, 40
USAF 1951 resolution target, 160–162

Varifocal lens, 87–92
Vignetting, 51, 186, 187
Virtual image, 6
Visual resolution, 123

Wax, tacky, 159, 160
Wide fields with stock lenses, 184–187
Wide screen movies, 83, 84
Wide-angle attachment, 66–68
Windows, exit and entrance, 57
Working f-number, 55

Zinc selenide, 149, 150
Zinc sulfide, 149, 150
Zoom lens, 87–92
  aperture stop, 88
  focusing of, 88

ABOUT THE AUTHOR

Warren J. Smith (Vista, CA), a chief scientist at Kaiser
Electro-Optics and also an independent consultant, is one of
the most widely known writers and educators in the field of
optical design. He is the author of *Modern Optical
Engineering* and *Modern Lens Design.*

Printed in the United States
3382